Kaushik Kumar, J. Paulo Davim (Eds.)

Hierarchical Composite Materials

Advanced Composites

Also of interest

Series: Advanced Composites
J. Paulo Davim (Ed.)
ISSN 2192-8983
Published titles in this series:
Vol. 7: *Green Composites* (2017) Ed. by Davim, J. Paulo
Vol. 6: *Wood Composites* (2017) Ed. by Aguilera, Alfredo/
Davim, J. Paulo
Vol. 5: *Ceramic Matrix Composites* (2016) Ed. by Davim, J. Paulo
Vol. 4: *Machinability of Fibre-Reinforced Plastics* (2015)
Ed. by Davim, J. Paulo
Vol. 3: *Metal Matrix Composites* (2014) Ed. by Davim, J. Paulo
Vol. 2: *Biomedical Composites* (2013) Ed. by Davim, J. Paulo
Vol. 1: *Nanocomposites* (2013) Ed. by Davim, J. Paulo/Charitidis,
Constantinos A.

Shape Memory Polymers
Theory and Application
Kalita; 2018
ISBN 978-3-11-056932-2, e-ISBN 978-3-11-057017-5

Nanocellulose
From Nature to High Performance Tailored Materials
Dufresne, 2017
ISBN 978-3-11-047848-8, e-ISBN 978-3-11-048041-2

Nanomaterials in Joining
Constantinos A. Charitidis (Ed.), 2015
ISBN 978-3-11-033960-4, e-ISBN 978-3-11-033972-7

Hierarchical Composite Materials

Materials, Manufacturing, Engineering

Edited by
Kaushik Kumar and J. Paulo Davim

DE GRUYTER

Editors
Dr. Kaushik Kumar
Birla Institute of Technology
Department of Mechanical Engineering
Mesra, Ranchi
Jharkhand 835215
India
kkumar@bitmesra.ac.in

Prof. Dr. J. Paulo Davim
University of Aveiro
Department of Mechanical Engineering
Campus Santiago
3810-193 Aveiro
Portugal
pdavim@ua.pt

ISBN 978-3-11-054400-8
e-ISBN (E-BOOK) 978-3-11-054510-4
e-ISBN (EPUB) 978-3-11-054402-2

Library of Congress Cataloging-in-Publication Data
Names: Kumar, K. (Kaushik), 1968- editor. | Davim, J. Paulo, editor.
Title: Hierarchical composite materials : materials, manufacturing,
engineering / edited by Kaushik Kumar and J. Paulo Davim.
Description: Berlin ; Boston : Walter de Gruyter, GmbH, [2018] | Series:
Advanced composites ; 8 | Includes bibliographical references and index.
Identifiers: LCCN 2018006866 | ISBN 9783110544008 (hardcover)
Subjects: LCSH: Composite materials. | Layer structure (Solids)
Classification: LCC TA418.9.C6 H52945 2018 | DDC 620.1/18–dc23 LC
record available at https://lccn.loc.gov/2018006866

Bibliographic information published by the Deutsche Nationalbibliothek
The Deutsche Nationalbibliothek lists this publication in the Deutsche Nationalbibliografie; detailed bibliographic data are available on the Internet at http://dnb.dnb.de.

Cover image: gettyimages/thinkstockphotos, Abalone Shell
Typesetting: Integra Software Services Pvt. Ltd.
Printing and binding: CPI books GmbH, Leck

www.degruyter.com

Preface

The editors are pleased to present the book *Hierarchical Composite Materials* under the book series *Advanced Composites*. Book title was chosen as it converges upcoming technologies in composite materials for the next decade. This book is a compilation of research in Hierarchical Composite Materials. In the present time, "Composite" is the buzzword for major disciplines and many scholars are working in these areas. This book provides insight for all researchers, academicians, postgraduates and senior undergraduate students working in the area.

Hierarchical materials can be natural and man-made. At the present level of sophistication, the idea of hierarchical material can be attributed to the physical properties of the material or the structure. Hence the idea of hierarchical structure can be the basis of the development of new microstructures that provide enhanced or more useful physical properties. The concept of functionally graded materials (FGMs) or composites comes from this part and forms an important constituent of this concept.

FGMs were introduced as a class of advanced materials composed of more than one continuous or discontinuous gradient in composition and microstructure. These materials have a flexibility in terms of the behavior as the property of one side of the FGMs differs from the other side. One side of the FGM structures could have higher mechanical strength and thermal resistance while the other side has opposite or different worth on the property. Position-dependent chemical composition, microstructure or atomic orders are among the factors that influence the property gradient in the materials. By controlling all these factors, the property gradient could be customized to meet any specific need within the best utilization of the composite components. FGM occurs in nature as teeth, bones and so on. Nature designed this material to meet its expected service requirement.

The concept of FGM was first considered in Japan in 1984 during a space plane project. That time the aim was to fabricate a space plane body's material with improved thermal resistance and mechanical properties by gradually changing (grading) compositions. Therefore, the space plane body is resistant to severe condition from temperature gap (about 1,000 °C gap) in between the inside and the outside.

As the products are still in development stage, challenges in developing innovative technique for FGMs preparation such as the adaptability for mass production and up-scaling, the process repeatability and reliability, the cost-effectiveness and the convenience level in controlling the quality can be achieved. Nowadays, researches on FGMs have been carried out intensively. Since the concept developed in aeronautics field in 1984, FGMs are also a concern in other fields such as industrial materials, optoelectronics, biomaterials, and energy materials. FGMs offer great promise in applications where the operating conditions are severe. Potential

https://doi.org/10.1515/9783110545104-201

applications include those structural and engineering uses that require combinations of incompatible functions such as refractoriness or hardness with toughness or chemical inertness with toughness. In aerospace and nuclear energy applications, reliability rather than cost is the a key issue. On the other hand, in applications such as cutting tools, machine parts and engine components, which require wear, heat, mechanical shock and corrosion resistance, the key issues are the cost/performance ratio and reliability. Examples of available commercial FGM products are high-performance cutting tools of tungsten carbide/cobalt-based FGMs and razor blades of an iron aluminide/stainless steel FGM.

The main objective of the book is the current status of hierarchical material, its development and application, and this targets audience such as academics, students, researchers and industry practitioners, engineers, research scientists/academicians involved in hierarchical materials.

The chapters in the book have been provided by researchers and academicians working in the field and have gained considerable success in the field.

Chapter 1 introduces the readers to an overview of the advancements in the fabrication processes for the FGMs. The chapter addresses the prospects and the challenges faced in using the different manufacturing processes highlighting relationship between the fabricated structures using different processes and the resulting properties. The chapter also investigates the applications of the key FGMs fabricated using different processing techniques. The chapter concludes with the future prospects of FGMs.

Chapter 2 indicates a review of manufacture of aluminium-based materials. These materials are becoming very popular due to their physical and mechanical characteristics, which makes them relevant to various applications. The addition of reinforcement materials with unique characteristics into aluminium produces aluminium composites with superior quality. Wear resistance, stiffness, strength and hardness are some of the improved properties obtained when reinforcement materials were added to the primary aluminium. This chapter presents some of the manufacturing processes of aluminium, its alloys and composites. The effects of reinforcements on aluminium composites from existing work and research direction on the fabrication of aluminium composite materials are discussed in this chapter.

Chapter 3 discusses about production of functionally graded composite materials using an advanced manufacturing method called additive manufacturing (AM) technology in general and laser additive manufacturing technologies in particular have been discussed. The technique includes selective laser melting/sintering and laser metal deposition/laser material deposition (LMD). Moreover, the recent research progresses are also presented. AM technology is an advanced manufacturing process used for the production of three-dimensional objects simply by adding materials in layers using the digital image of the component.

Chapter 4 illustrates fabrication of stainless steel-based FGM by laser metal deposition. This chapter starts with an overview of the concept of FGMs in terms and its history. Finally, a study on the microstructure and microhardness of laser deposited compositionally graded 316 L/17-4PH has been presented.

In Chapter 5, natural fiber-reinforced composites (NFCs) as FGMs for advanced applications have been taken up. The characterization of NFCs has led to a new era for the utilization of its properties in multiple tasks and a variety of applications. This chapter aimed at evaluation of NFCs to be as FGMs for the advanced applications considering methodologies for material selection. The material under scanner was natural fiber-reinforced polymer or co-polymer functionally graded composites. The composite has shown respectable results for a considerable number of applications, for example, automotive industry, constructions and renewable energy production.

Chapter 6 works with the performance of functionally graded longitudinal fins having varying geometry. A comparison of the performance of the parabolic longitudinal fins having fixed aspect ratio with longitudinal transition in thermal conductivity approximated by means of a power series is reported by solving the second-order governing differential equation. The fin is insulated at the tip with a predefined temperature at the base. A parametric study was then carried out by controlling the geometry parameters and grading parameters. The solution has been obtained using *bvp4 c* subroutine, a technique of solving boundary value problems using continuous piecewise polynomial that adjusts the boundary conditions and is validated with benchmark results. Further, the results have been reported for combinations of geometry and grading parameters and are presented in graphical and tabular forms. The results give a substantial insight into the behavior of longitudinal fins and can be used as design data.

In Chapter 7, analytical approach for transient response of functionally graded rectangular plates including the higher-order shear deformation effects has been elaborated. The governing equations account for higher-order shear deformation in functionally graded plate (FGP). The material properties of the plate were assumed to vary continuously in the thickness direction according to the power-law form. Analytical solution based on state variable method is presented for both free and forced vibrations of simply supported functionally graded rectangular plates. The results obtained using analytical approach were compared with those obtained by finite element method. The effects of power-law exponent index on the behavior of FGP are also studied.

Chapter 8 applies fuzzy approach in assessment of frequency response function (FRF) analysis of FGPs. The fuzziness has been considered due to variability in material properties corresponding to the various α-cuts. The power law was implied for characterizing the material modeling, and a parametric study was carried out to observe the effect of location of drive point and cross points on uncertain bounds of FRF with respect to crisp values.

Chapter 9 performs experimental analysis of FGM product. In this chapter, the transmission efficiency of glass fiber-reinforced homogeneous polyamide 66 (PA66) composite gears and FGM-based PA66 gears has been investigated and compared. Homogeneous gears were fabricated by conventional technique, whereas FGM-based PA66 gears were fabricated using a novel manufacturing system. Experiments were performed at various speed and torque combinations for 0.2 million cycles. Results show that FGM gears have higher transmission efficiency as compared to neat and homogeneous PA66 gears.

The last chapter of the book, i.e., Chapter 10, talks about the performance of thick cylinders made of FGM. In this chapter, variational principle was employed to investigate the stress and deformation states for the thick cylinder. Based on Galerkin's error minimization principle upon a series approximation, the solution of the governing equation was obtained. The property of FGMs was decided on exponential spatial variation. Stress and displacement along the cylinder radius was studied and plotted. The results obtained were validated with standard experimental data and were found to be in good agreement.

First and foremost, the editors would like to thank God. In the process of putting this book together, it was realized how true this gift of writing is for anyone. You have given the power to believe in passion, hard work and pursue dreams. This could never have been done without the faith in You, the Almighty. The editors would also like to thank all the chapter contributors, reviewers, the editorial board members, project development editor and the complete team of publisher Verlag Walter de Gruyter GmbH for their availability to work on this editorial project.

Kaushik Kumar
J. Paulo Davim

Contents

Preface — V
List of Contributing Authors — XIII
Editors' Biography — XV

Divya Zindani, Kaushik Kumar, and J. Paulo Davim
1 Fabrication of functionally graded materials: A review — 1
1.1 Introduction — **1**
1.2 Solid-based processes — **2**
1.2.1 Powder metallurgy — **2**
1.2.2 Spark plasma sintering — **4**
1.3 Liquid-based processes — **5**
1.3.1 Electrochemical gradation — **5**
1.3.2 Sedimentation — **5**
1.3.3 Direct solidification — **5**
1.3.4 Laser deposition — **6**
1.3.5 Chemical solution deposition — **6**
1.3.6 Electrophoretic deposition — **6**
1.3.7 Slip casting — **7**
1.3.8 Gel casting — **7**
1.3.9 Centrifugal casting — **8**
1.3.10 Combustion — **9**
1.4 Gas-based processes — **10**
1.4.1 Surface reaction process — **10**
1.4.2 Chemical vapour deposition/infiltration — **10**
1.4.3 Thermal spray — **11**
1.5 Conclusion — **13**
References — **14**

Kamardeen O. Abdulrahman, Esther T. Akinlabi, and Rasheedat M. Mahamood
2 Manufacturing of aluminium composite materials: A review — 21
2.1 Introduction — **21**
2.2 Casting processes for aluminium and its alloys — **23**
2.2.1 Casting processes — **24**
2.3 Other manufacturing process of aluminium and its alloys — **26**
2.3.1 Forging process for aluminium and its alloys — **27**
2.4 Manufacturing of aluminium composite materials — **29**
2.4.1 Stir casting method — **30**
2.4.2 Powder metallurgy — **33**

2.5 Functionally graded materials of aluminium composites and research direction of manufacturing of these composite materials — **34**
2.6 Summary — **38**
References — **38**

R. M. Mahamood, E. T. Akinlabi, G. M. Owolabi, and K. O. Abdulrahman
3 Advanced manufacturing of compositionally graded composite materials: An overview — 41
3.1 Introduction — **41**
3.2 Production of functionally graded materials with selective laser sintering/melting — **43**
3.3 Manufacturing of functionally graded material using laser metal deposition process — **46**
3.4 Future research direction — **50**
3.5 Summary — **51**
References — **52**

Abiodun Bayode, Esther Titilayo Akinlabi, and Sisa Pityana
4 Fabrication of stainless steel-based FGM by laser metal deposition — 55
4.1 Introduction — **55**
4.1.1 Development of FGM — **56**
4.1.2 Applications of FGMs — **56**
4.1.3 Classification of FGMs — **58**
4.1.4 Manufacturing techniques — **58**
4.1.5 AM technologies — **59**
4.1.6 Process parameters — **60**
4.1.7 Laser-deposited FGM — **62**
4.2 Experimental — **63**
4.2.1 Materials — **63**
4.2.2 Experimental set-up — **64**
4.2.3 Processing — **65**
4.2.4 Specimen characterization — **66**
4.3 Results and Discussion — **66**
4.3.1 Macrostructure — **66**
4.3.2 Microstructure — **67**
4.3.3 Microhardness — **70**
4.4 Conclusions — **70**
References — **71**

M. H. Alaaeddin, S. M. Sapuan, M. Z. M. Yusoff, E. S. Zainudin, and Faris M. AL-Oqla
5 Natural fiber composites as functionally graded materials for advanced applications — 73
5.1 Introduction — 73
5.2 NFCs for potential use as FGM — 74
5.3 Material selection in NFCs as FGMs — 75
5.4 Methods of materials selection that can be used to select NFCs as FGMs — 76
5.5 Computerized material databases — 77
5.6 Functionally graded materials — 79
5.7 FGM modeling — 81
5.8 Conclusions — 87
References — 87

Savita K. Subramaniam, Vivek Kumar Gaba, and Shubhankar Bhowmick
6 Temperature distribution in functionally graded longitudinal fins of varying geometry — 90
6.1 Introduction — 90
6.2 Mathematical formulation — 91
6.3 Results and discussions — 94
6.4 Conclusion — 97
References — 97

Ta Duy Hien
7 Analytic approach for transient response of functionally graded rectangular plates including the higher-order shear deformation effects — 99
7.1 Introduction — 99
7.2 Governing equations and solution procedures — 100
7.2.1 The plate model — 100
7.2.2 Governing equations — 102
7.2.3 Solution procedures — 105
7.3 Results and discussions — 108
7.3.1 Comparison of the present approach with ABAQUS software — 109
7.3.2 Transient response with various loadings — 112
7.4 Conclusions — 116
References — 116

P. K. Karsh, T. Mukhopadhyay, and S. Dey
8 Fuzzy-based frequency response function analysis of functionally graded plates — 119
8.1 Introduction — 119
8.2 Background — 120
8.3 Governing equations — 123
8.4 Fuzzy-based approach — 124
8.5 Results and discussion — 129
8.6 Conclusions — 136
References — 136

Akant Kumar Singh, Siddhartha, and Prashant Kumar Singh
9 Transmission efficiency of glass fiber-filled functionally graded material-based PA66 composite spur gears — 140
9.1 Introduction — 140
9.2 Experimental details — 142
9.2.1 Gear fabrication — 142
9.2.2 Polymer gear test rig — 143
9.2.3 Methodology to verify the gradation in FGM specimen — 144
9.2.4 Dynamic mechanical analysis and transmission efficiency — 146
9.3 Results and discussions — 147
9.3.1 Experimental verification of gradation in FGM gears by ignition loss test — 147
9.3.2 Experimental verification of gradation in FGM gears by hardness measurement — 147
9.3.3 Dynamic mechanical analysis — 148
9.3.4 Effect of speed and torque on the transmission efficiency of PA66-based homogeneous and FGM gears — 150
9.4 Conclusions — 153
References — 153

Lakshman Sondhi, Subhashis Sanyal, Kashinath Saha, and Shubhankar Bhowmick
10 Approximate solution of functionally graded thick cylinders — 156
10.1 Introduction and literature review — 156
10.2 Mathematical formulation — 157
10.3 Results and discussions — 159
10.4 Conclusions — 163
References — 163

Index — 165

List of Contributing Authors

Kamardeen O. Abdulrahman
Department of Mechanical Engineering Science, University of Johannesburg, Auckland Park Kingsway Campus, Johannesburg 2006, South Africa
olajideabdulrahman@gmail.com

Alaaeddin M. H. Abed
Department of Mechanical and Manufacturing Engineering, Faculty of Engineering, Universiti Putra Malaysia, 43400 UPM Serdang, Selangor, Malaysia
alaaeddinabed@gmail.com

Esther Titilayo Akinlabi
Department of Mechanical Engineering Science, University of Johannesburg, Johannesburg, South Africa
etakinlabi@uj.ac.za

Faris M. AL-Oqla
Laboratory of Bio-Composite Technology, Institute of Tropical Forestry and Forest Products, Universiti Putra Malaysia, 43400 UPM Serdang, Selangor, Malaysia
farisv9@gmail.com

Abiodun Bayode
Department of Mechanical Engineering Science, University of Johannesburg, Johannesburg, South Africa
reachabeyy@gmail.com

Shubhankar Bhowmick
Department of Mechanical Engineering, National Institute of Technology Raipur 492010, India
sbhowmick.mech@nitrr.ac.in

J. Paulo Davim
University of Aveiro
Mechanical Engineering Department
Campus Santiago
3810–193 Aveiro
Portugal
pdavim@ua.pt

S. Dey
Mechanical Engineering Department, National Institute of Technology, Silchar, Assam, India
infodrsudip@gmail.com

Ta Duy Hien
Department of Civil Engineering, University of Transport and Communications, Ha Noi, Vietnam
tdhien@utc.edu.vn

P. K. Karsh
Mechanical Engineering Department, National Institute of Technology, Silchar, Assam, India
pradeepkarsh@gmail.com

Kaushik Kumar
Department of Mechanical Engineering, Birla Institute of Technology, Ranchi, India
kkumar@bitmesra.ac.in

Vivek Kumar Gaba
Department of Mechanical Engineering, National Institute of Technology, Raipur, India
vgaba.mech@nitrr.ac.in

Akant Kumar Singh
Department of Mechanical Engineering, NIT, Hamirpur, Himachal Pradesh, India
akant.nith@gmail.com

Prashant Kumar Singh
Department of Mechanical Engineering, NIT, Hamirpur, Himachal Pradesh, India
nithprashant@gmail.com

Rasheedat M. Mahamood
Department of Mechanical Engineering, University of Ilorin, Ilorin, Nigeria
mahamoodmr2009@gmail.com

T. Mukhopadhyay
Department of Engineering Science, University of Oxford, Oxford, UK
mukhopadhyay.mail@gmail.com

https://doi.org/10.1515/9783110545104-202

M.G. Owolabi
Department of Mechanical Engineering,
Howard University, Washington, DC, USA
gbadebo.owolabi@howard.edu

Sisa Pityana
National Laser Centre, The Council of Scientific and Industrial research
Pretoria, South Africa
spityana@csir.co.za

Kashinath Saha
Department of Mechanical Engineering,
Jadavpur University, Kolkata 700032, India
kashinathsaha@gmail.com

Subhashis Sanyal
Department of Mechanical Engineering, NIT,
Raipur 492010, India
ssanyal.mech@nitrr.ac.in

S. M. Sapuan
Department of Mechanical and Manufacturing Engineering, Faculty of Engineering, Universiti Putra Malaysia, 43400 UPM Serdang,
Selangor, Malaysia
sapuan@upm.edu.my

Siddhartha
Department of Mechanical Engineering, NIT,
Hamirpur, Himachal Pradesh, India
sid.fgm@gmail.com

Lakshman Sondhi
Department of Mechanical Engineering, SSGI,
Bhilai, Chhattisgarh, India
lsondhii@gmail.com

Savita K. Subramaniam
Department of Mechanical Engineering,
National Institute of Technology, Raipur, India
ks.savita7@gmail.com

Mohd Zuhri Mohamed Yusoff
Department of Mechanical and Manufacturing Engineering, Faculty of Engineering, Universiti Putra Malaysia, 43400 UPM Serdang,
Selangor, Malaysia
zuhri@upm.edu.my

E. S. Zainudin
Department of Mechanical Engineering,
Faculty of Engineering, The Hashemite University, 13133 Zarqa, Jordan
edisyam@upm.edu.my

Divya Zindani
Department of Mechanical Engineering,
National Institute of Technology, Silchar,
Assam, India
divyazindani@gmail.com

Editors' Biography

Kaushik Kumar, B.Tech. (Mechanical Engineering, REC (Now NIT), Warangal), MBA (Marketing, IGNOU) and Ph.D. (Engineering, Jadavpur University), is presently an associate professor in the Department of Mechanical Engineering, Birla Institute of Technology, Mesra, Ranchi, India. He has 15 years of experience in teaching and research and over 11 years of industrial experience in a manufacturing unit having a global reputation. His areas of teaching and research interest are quality management systems, optimization, non-conventional machining, CAD/CAM and rapid prototyping and composites. He has nine patents, 14 books, six edited books, 35 book chapters, 120 international journal publications, 18 international and eight national conference publications to his credit. He is on the editorial board and review panel of seven reputed international journals and one national journal. He has been felicitated with many awards and honours.

J. Paulo Davim received his Ph.D. in mechanical engineering in 1997, M.Sc. in mechanical engineering (materials and manufacturing processes) in 1991, Licentiate degree (5 years) in mechanical engineering in 1986 from the University of Porto (FEUP), the Aggregate title from the University of Coimbra in 2005 and D.Sc. from London Metropolitan University in 2013. He is EUR ING by FEANI and senior chartered engineer by the Portuguese Institution of Engineers with MBA and specialist title in engineering and industrial management. Currently, he is professor at the Department of Mechanical Engineering of the University of Aveiro. He has more than 30 years of teaching and research experience in manufacturing, materials and mechanical engineering with special emphasis in machining and tribology. Recently, he has also interest in management/industrial engineering and higher education for sustainability/engineering education. He has received several scientific awards. He has worked as evaluator of projects for international research agencies as well as examiner of Ph.D. thesis for many universities. He is the editor in chief of several international journals, guest editor of journals, editor of books, series editor of books and scientific advisor for many international journals and conferences. At present, he is an editorial board member of 25 international journals and acts as reviewer for more than 80 prestigious Web of Science journals. In addition, he has also published as editor (and co-editor) for more than 100 books and as author (and co-author) for more than 10 books, 80 book chapters and 400 articles in journals and conferences (more than 200 articles in journals indexed in Web of Science/h-index 45+ and SCOPUS/h-index 52+).

https://doi.org/10.1515/9783110545104-203

Divya Zindani, Kaushik Kumar, and J. Paulo Davim

1 Fabrication of functionally graded materials: A review

Abstract: The present work presents an overview of the advancements in the fabrication processes for the functionally graded materials (FGMs). The work addresses the prospects and the challenges faced in using the different manufacturing processes. The relationship between the fabricated structures using different processes and the resulting properties is discussed. Further, the work discusses the applications of the key FGMs fabricated using the different processing techniques. The chapter concludes with the future prospects of FGMs. The authors feel that the review is a comprehensive one and give an insight for the scientific community in the field of FGMs.

Keywords: functionally graded materials, fabrication techniques, solid-based processes, liquid-based process, gaseous processes, material properties

1.1 Introduction

In today's world of innovation, a large number of innovative materials have been developed by the scientific community. The economic and environmental importance of these materials has been realized by the scientists and engineers of the modern era [1]. One of the main classes of such materials is known as the functionally graded materials (FGMs) that have an in-depth graded composition thereby lending it for tailored properties [2]. FGMs such as bamboo have been in existence for thousands of years and were used for construction and decorative purposes [3]. The theoretical implications of the materials with graded composition were investigated by Shen and Bever [4]. However, development of these materials was delayed for over a decade due to the limitation on the fabrication processes available [5]. The scientific terminology for these materials, "functionally grade materials", was coined in the year 1984 where they were used for the fabrication of thermal barriers in Japan [2, 6]. Since then, the applications of FGMs have risen and are being used in numerous industries: aerospace, nuclear and bioengineering industries. As such the number of technical reports on FGMs has also escalated in the past 20 years.

The specific spatial distribution of the constituent phases such as ceramics, polymers and metals in a continuing manner and with slight variations results in FGMs. One of the major advantages of FGMs over other composites is the tailored structural properties and the morphological properties [7, 8]. The schematic of FGM is shown in Figure 1.1, which depicts gradation in microscopic properties such as wear resistance, thermal

https://doi.org/10.1515/9783110545104-001

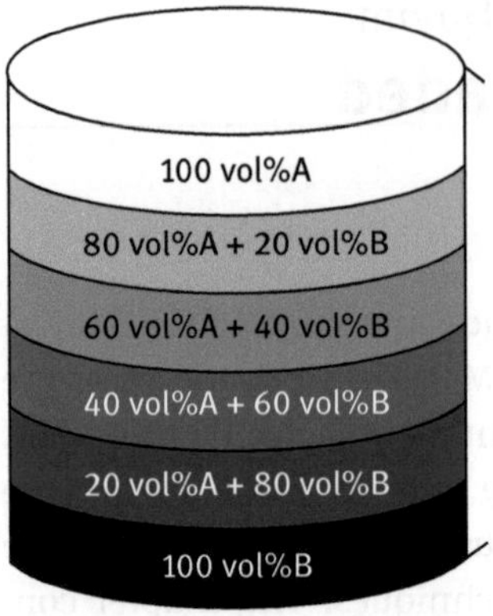

Figure 1.1: A schematic of functionally graded materials. A denotes material A; B denotes material B.

conductivity, hardness, specific heat in case of the thermal barrier coatings used for furnace liners, electromagnetic sensors and piezoelectric actuators [9 – 12].

To produce composite materials with compositional gradients a number of methods are used. These methods can be categorized into liquid phase, gas-based and solid phase methods [13]. Some of the examples of gas-based methods include plasma spraying, ion plating and chemical vapour deposition (CVD) technique (CGD) [14, 15]. The control of the production system and the ratio of the phase's reaction determine the gradient composition of the deposited products in case of vapour deposition systems. The compositionally gradient composites obtained by the liquid phase methods are popular for coating applications. The final properties of the FGM depend on the type of processing technique used. The two main steps in fabrication of FGMs are the spatial gradation of the components in nonhomogeneous manner and consolidation which involves the setting of the graded structure on a substrate.

FGM lends itself into a large number of applications. Al_2O_3/Y-ZrO_2 fabricated FGMs are used in medical applications such as for hip or knee prosthesis [16–18]. The graded dielectric is used in the energy sector [19], high-current connectors, sensors and capacitors [20–22].

Therefore, the present work provides an overview of the advancements in the research of FGMs. The chapter presents a review on the key opportunities in research relating to the fabrication processes for the FGMs. The future prospects of FGMs in different technological and scientific fields are discussed.

1.2 Solid-based processes

1.2.1 Powder metallurgy

The powder metallurgy process of fabrication involves preparation of powdered mixture, stacking and sintering. One of the easiest methods to obtain green compact is step-by-step stacking method. For fabrication of thin-film gradient structure, spray forming processes and the other continuous powder stacking methods are used. However, the stacking method comes with certain limitations such as warping, formation of frustum, initiation and propagation of crack as well as occurrence of lamination due to the distribution of particles in an unpredictable manner. Addition of activation phase and controlling the process of sintering are recommended to achieve the best results for FGMs. The imbalance of the sintering process can be avoided by controlling the particle size, establishing the best possible mixing condition and application of the hot pressing at the end of the sintering process [2, 23, 24]. FGMs produced using the powder metallurgy techniques are produced with good control of microstructure and chemical composition [25].

An investigation on the fabrication of hydroxyapatite-Ti FGM using the powder methodology process was carried out by Chenglin et al. [26]. A gradual change in the microstructure of the (HAp)-Ti FGM was reported with the gradual change in the chemical composition of the FGM. The hardness and the bonding strength of FGM were affected as a result of this gradual variation for the different zones. The decomposition of HA to $Ca_4P_2O_9$ and α-$Ca_3(PO_4)_2$ at sintering temperature of 1,100 °C was reported to be catalysed by the presence of Ti. Further, the fracture toughness and the bending strength were found to have a direct relationship with the Ti content, being highest and comparable to that of human bone strength in the regions with pure Ti. Hard tissue implant is one of the main applications of (HAp)-Ti FGMs. Mullite/Mo FGM was developed by Jin et al. [25] using the powder metallurgy process. The system had components with similar thermal expansion coefficients and therefore aided in the reduction of residual thermal stresses due to expansion mismatch. The thermal behaviour of the fabricated FGMs was found to follow the Reuss model and the mechanical properties followed the Vogit model. Ma and Tan [27] studied the effect of adding SiC powders on the structural and mechanical properties of Ti/TiB_2 FGM. A higher toughness and flexural strength was reported for the Ti/TiB_2 FGMs using SiC powders as sintering aid.

Pressure-assisted densification method was used for the fabrication of Ti-TiB FGMs for ballistic application [28]. The drawbacks of residual stress were addressed by the development of Ti-TiB FGMs. The mechanical properties such as modulus of elasticity, the thermal coefficient of expansion and ballistic performance were reported to be strongly influenced. On the similar track the ballistic performance of SiC-AA7075 FGM, fabricated using the powder metallurgy technique, was investigated by Úbeyli et al. [29]. The fabricated FGMs of different thickness were tested using armour-piercing projectile for the ballistic performance. An enhanced ballistic performance was reported for specimen with thickness above 25 mm. A brittle deformation was reported in the fabricated FGMs with microcracks being observed under the loading zone for the sample.

Polymeric functionally graded nanocomposites were fabricated using the powder metallurgy process from phenolic resin and nanofibres and were investigated for the viscoelastic properties and thermomechanical properties [30–33]. A maximum improvement of around 97% in the thermomechanical properties and 34.7% for the creep resistance was observed for the composites with highest percentage of graphite in the bottom and top layers while lowest percentage at the centre.

1.2.2 Spark plasma sintering

One of the most widely used solid-phase processes of FGM fabrication is the spark plasma sintering (SPS) [34–36]. The principle of the SPS process could be found in the bricks of the Mesopotamia civilization made from the heated clay. A sintering process for fabrication of gold jewellery and decorative tools from modification of properties of ceramics and other metals was applied, respectively, by Incas in America and inhabitants of Egypt and Persia. The durability and usefulness of fabricated materials were enhanced with the further noticeable advances in China, Middle East and Europe. It was in 1910 that a process similar to standard SPS was first applied in Germany. Further developments were made in the process developed in the form of patent published on resistance sintering method by Tayler in 1933 [37] for sheet metals, which were then extended for the sintering of brass and bronze by Cremer [38]. The focus of the scientific community on the phenomenon of mass and solid-state transport in the materials was expanded, and further developments in this field were taken up by German, Kingery and others [39–41]. With further advancements, the range of materials that can be dealt with the SPS process was expanded further.

SPS technique lends itself to a large number of applications such as fabrication of high-wear resistance material and sandblasting nozzles, besides being used for the fabrication of intermetallic compounds, FGMs, metal matric composites, nanoproducts and fibre-reinforced metal matrix composites [42]. SPS process has the process capability to obtain different phase melting temperature and is therefore suitable for the fabrication of FGMs. It is relatively easy to activate and purify the powders in SPS technique in comparison to other conventional techniques. The diffusion process is enhanced because of the migration of ions at high speed on application of high-intensity current. Localized regions with high temperature can be achieved by discharging between the powders. The softer and molten components of the system then fill the gap between the particles [43, 44].

Yuan et al. [34] fabricated the ZrB_2 FGM using the SPS-reactive synthesis technique. The FGM was fabricated using ZrO_2 and B_4C powders and was investigated for the chemical reactions between the two powders at different temperatures. The best sintering conditions were obtained, which consisted of the holding time and the processing temperature. Further, it was established that by using a high sintering temperature for the minimum holding time, the surface morphology could be modified [45]. The different FGM systems were investigated for the best possible SPS process parameters by different researchers [46–56].

An insight into current-activated tip-based sintering was presented by El-Desouky et al. [57] and established that a direct relationship between FGM size and the green density existed. Yongming et al. [49] studied the Al_2O_3/Ti_3SiC_2 FGM system and observed an increase in electrical conductivity of the FGM with the increased weight percentage of Ti_3SiC_2. Further, increase in the hardness of the FGM was also found to be in direct relation to that of the Ti_3SiC_2 content. A bio-FGM, hydroxyapatite-alumina-zirconia was fabricated and was investigated for the structural integrity by Afzal et al. [58].

1.3 Liquid-based processes

1.3.1 Electrochemical gradation

The method can be used to produce graded structures with the use of wide range of ceramics and metallic. The W/Cu FGM system was fabricated using the electrochemical gradation technique by Jedamzik et al. [59]. They were able to produce the required chemical gradient for the FGM and reported that the shape and the morphological structure was affected by some of the process parameters such as electrolytic porosity, resistivity, density and also its geometry.

1.3.2 Sedimentation

Particle-reinforced composites are especially produced using the process of sedimentation. The sedimentation process is based on the difference in density between the particle and the matrix. A graded structure can be produced with the proper use of the gravitational force. A wide range of functional graded polymer matrix composites and metal matrix composites can be produced by focusing on the morphology, thermal cooling gradient and size and mass diffusion of the constituent phases [42, 60]. Functionally graded polyester was fabricated by Quadrini et al. [61] and was investigated for the mechanical properties.

1.3.3 Direct solidification

The method of direct solidification lends itself for a large number of materials especially for ceramics and metallic-based systems. One of the important aspects of this method is to provide single crystal for the purpose of hi-tech applications. Organic polymeric blends were used in the fabrication of poly-ε-caprolactone)/thiodiphenol (PCL/TDL) FGM by employing the Bridgman method [62]. Increase in PCL/TDL hydrogen bonding was reported with decrease in temperature.

1.3.4 Laser deposition

The laser deposition technique works to deposit ceramic or metal powder onto a substrate using high-intensity laser. The method is suitable for cases in which there is a difference in melting temperature of the matrix and the material used for reinforcements. Dense net structures with enhanced mechanical properties can be achieved using the laser deposition technique [63, 64]. Functionally graded Ti-Al was fabricated by Shishkovsky et al. [65] using the laser deposition method and investigated the physical behaviour for the crack initiation and propagation. The Ti-Al FGMs are used in aerospace and nuclear fields owing to the higher value of modulus of elasticity, excellent mechanical properties and density which can be obtained by optimizing the various process parameters. Nickel-based functionally graded coatings were fabricated using the laser deposition technique by Wilson and Shin [63]. It was reported that a more refined structure could be obtained with increase in the TiC powders to Inconel 690 powders.

1.3.5 Chemical solution deposition

Different aspects of chemical solution deposition (CSD) techniques have been discussed in a review report by Schwartz et al. [66]. The aspects in relation to the electronic oxide films have been discussed. Further, techniques such as metalorganic deposition, hybrid and sol–gel techniques have also been discussed in detail. The CSD technique is simple and inexpensive for the processing of thin film which makes use of a liquid precursor dissolved in a liquid solvent. CSD process has been used for the application of TiO coatings on the $Si/SiO_z/TiO_x/Pt$-substrate by Slowak et al. [21]. It was demonstrated that low-cost dielectric thin-film capacitor can be obtained with the careful selection of the process parameters.

1.3.6 Electrophoretic deposition

The observation of clay particles in water under the influence of electric field by the Russian scientists gave recognition to the electrophoretic deposition (EPD) technique. The EPD technique consists of different industrial processes such as anodic and cathodic electrodeposition, electrophoretic coatings, electrophoretic painting and electrocoating. Deposition of particles of thoria on the platinum electrode was the first experimental work carried out by the scientists from the USA in this regard. The EPD technique gained popularity in the 1980s when it was used for fabrication of ceramic tools. Some of the major advantages of the EPD technique are process simplicity, short processing time and no need for burnout of blinder [67].

The mechanical behaviour of WC/Co and ZrO_2/Al_2O_3 FGM systems was investigated by Put et al. [68] under the impact of EPD processing technique. The variation in hardness value was found to be ranging between 13.5 GPa at the core to 19.5 GPa at the edges for the ZrO_2/Al_2O_3 FGM system whereas for the WC/Co FGM system the hardness value ranged between 21 and 9 GPa from the core to edges. In another work by Farnoush et al. [69, 70], the Ti-6Al-4V substrate was deposited with the HA-TiO_2 coatings.

1.3.7 Slip casting

Katayama et al. [71] fabricated the AL_2O_3-W FGM system using the technique of slip casting with subsequent sintering at a temperature of 1,800 °C. An increase in the dispersibility of the raw materials was reported with higher percentage of dispersant. Further, a gradual increase in the grain size of W was also reported with the changing layer composition. Functionally graded alumina composites were fabricated using the slip casting technique and was studied for the heat treatment process by Andertovà et al. [72]. Layers with different percentages of porosity were obtained.

1.3.8 Gel casting

One of the widely used techniques for producing complex shaped ceramics with dense and porous structure is the gel casting technique. The gel casting technique is the colloidal processing technique that has the advantage of high green capacity, higher yield with low cost and shorter forming time. Scientists at the Oak Ridge National Laboratory were the first to carry out scientific work in this field [73]. Polymerization of monomers in the suspension solution is the key working phenomenon for the production of green products having high strength and demouldablity characteristics. The polymerization phenomenon takes place via free radical initiator [74, 75]. Gel casting technique has been used widely for producing FGMs and piezoelectric ceramic-polymer composites for ultrasound devices [76]. During the process, a castable slurry fluid is formed using mixture of ceramic powders and a liquid solution containing cross-linker, monomer, catalyst and a free radical initiator. The castable slurry is then poured into mould and then the polymerization process takes its course to form a polymer–water gel. The gel from the mould is removed, dried and sintered to terminate the gel casting process [73].

The Al_2O_3/ZrO_2 FGM system was fabricated by dip coating the substrates prepared by the gel casting technique by Park et al. [77] and was characterized for the mechanical and the microstructural behaviour. A smooth variation in the hardness was observed from the substrate to the outer layer.

1.3.9 Centrifugal casting

In this fabrication process for metallic FGMs, the reinforcement phase is mixed with the molten metal to form a uniform mixture. A chemical composition with the desired gradient is obtained by segregating the liquid and the reinforcement particles using the centrifugal/gravitational forces. This gradient can be maintained by controlling the solidification process [78]. The centrifugal casting process can be categorized into two main types depending upon the difference in processing and master alloy temperature. One of the categories is referred to as centrifugal in situ technique in which case the processing temperature is higher than that for the master alloy temperature. The other technique being referred to as centrifugal solid-particle temperature, where the master alloy experiences higher temperatures than the processing temperatures [79]. A mathematical approach for explaining casting process of metal matrix composites has been provided for better understanding of the mechanism of bonding property and diffusion mechanism [80, 81].

A number of metal-ceramics and metallic-intermetallic FGM systems have been fabricated using the centrifugal casting process [9, 82, 83]. One of the major challenging tasks in the centrifugal casting process is to achieve the homogeneous distribution of the second phase and hence the control of the final product with high precision. The gradient in composition is affected by the cooling and heating control systems as well as the rotating speed of the mould. The researchers have however adopted the centrifugal casting technique on a wide scale because of the numerous limitations associated with the other methods such as that of forging and powder metallurgy. Impact of the key process parameters such as moulding temperature, speed and pouring temperature has been investigated on the TiAl systems by Fu et al. [84]. It was reported that the solidification process was aided by decreasing the radiation effect from the neighbourhood. This resulted in elimination of the porosity in the modified design. The principles of fabricating FGMs using the casting process were studied by Fukui [85]. Rings were casted using conundrum and plaster mixture as introducer for different volume fraction and processing time, and the mathematical equation showing the relationship between gradient of powders with volume fraction and processing time was proposed. Fabrication of aluminium–silicon carbide FGM system was done by Rajan et al. [86] using the centrifugal casting method. The outer circumference of the Al2124 and Al-A356 was reported to consist of maximum percentage of SiC particulates ranging between 40% and 45%. The hardness of the Al2124-SiC and the Al-A356 FGMMCs was reported to be 155 and 145 Brinell, respectively, which was achieved after addition of extra reinforcements and heat treatment. The Al-SiC FGMMC systems obtained had superb hardness, strength and wear resistance.

The mechanical and microstructural behaviour of Al359/SiC_p FGM system was investigated for the SiC particles by Rodrıguez-Castro et al. [87]. It was reported that the content of the SiC particulates and the modulus of elasticity changed continuously by employing the centrifugal casting method. Increased tensile strength was observed for volume fraction ranging from 0.2 to 0.3 SiC; however, a marked decrease in the Ultimate tensile strength (UTS) was reported for the SiC volume fractions ranging from 0.3 to 0.4. Further with increasing SiC volume fractions, the stiffness and modulus of elasticity also increased.

The feasibility of centrifugal casting process to fabricate nickel-aluminide/steel clad pipe was explored by Watanebe et al. [88]. The coating area was found to have an inverse relation with the angle of the mould axis.

1.3.10 Combustion

The first work was published by Booth on combustion synthesis in 1953. The cost-effectiveness of the combustion method for the fabrication of the materials can be realized through the scientific reporting made by Merzhanov and Varma [89, 90]. Self-propagating high-temperature synthesis (SHS) is often the terminology used by the scientific community. The main principle of the SHS process is the exothermic chemical reaction between the reactive raw materials. SHS process has gained a wide range of acceptability among the industries involved in nanomaterials, ceramics and catalysts because of the simplicity of the process and the cost-effectiveness [91, 92].

Centrifugal-assisted SHS technique has been employed for the fabrication of the titanium carbide aluminide-alumina-iron FGM rings [91]. Enhanced mechanical and structural changes were reported to be experienced by Ti_3Al and TiAl in the TiC-Fe-Al_2O_3FGM system mainly because of the centrifugal forces and the high temperature of processing. TiC-Ni FGM system was fabricated using reactive compaction synthesis, and changes in structural and mechanical behaviour of the fabricated FGM were examined by Zhang et al. [93]. The relationship between nickel content and the physicomechanical properties was established. Further, it was reported that an optimum content of nickel is required for achieving the optimal mechanical properties. An initial increase in the properties such as hardness, fracture toughness and flexural strength was reported with the increasing Ni content. With the Ni percentage of 20 and 30 wt%, the hardness and the flexural strength reached their maximum values, respectively. However, a decreasing trend in the hardness and flexural strength was reported beyond further increase in Ni content. The fabrication process of TiC-Cu FGM system using the combustion process was studied by Shon and Munir [94]. The performance of the FGM system was compared for the two different processing techniques used: ignition using radiative heating and electrothermal combustion. Lower porosity was observed for the samples fabricated using the latter technique. The kinetics of processing for the NiTi-TiC_x FGMs fabricated using the combustion process was studied by Moore [95]. A high porosity with small pores for the layers with a higher percentage of TiC_x particles was reported, whereas equal porosity was reported for higher concentration of NiTi but with larger pore size.

1.4 Gas-based processes

1.4.1 Surface reaction process

Nitriding and carburizing are some of the surface reaction processes that are being employed for the treatment of steels and other metals [96–98]. Different surface treatment techniques were reviewed by Zhecheva et al. [99] for the titanium alloys such as plasma nitriding, ion nitriding and gas and laser nitriding. The conditions for the formation of the nitride layers as well as the kinetics of the nitriding were also discussed. They also examined the impact on the formation of new phases of the processing parameters such as time and temperature. The CoMo carbide bimetallic was fabricated using the carbothermal hydrogen reduction process by Liang et al. [100]. The method forms reaction products through the process of diffusion of the reactive gas into the substrate. The diffusion bonding process was employed for the joining of stainless steel and Wc-Co/Ni FGM by Feng et al. [101]. The joining strength was found to be dependent on the bonding temperature and the holding time. WC-Co/Ni FGM composites were fabricated with different WC-Co transition layer and Ni transitional layer by Chen et al. [102]. An increase in the hardness value was reported to increase in the direction from Ni layer to WC-Co layer. The mechanism of carbon diffusion bonding was analysed for WC-Co system by Guo et al. [103]. The kinetics of the metal binder during the liquid transfer was analysed. A thicker cobalt gradient was obtained with longer carburizing time and larger ratio of P_{CH4}/P_{H2}^2.

1.4.2 Chemical vapour deposition/infiltration

A patent on reduction process of WCl_6 by H_2 from de Lodyguine in 1893 [104] was the first instance where CVD process was utilized. This was utilized for the deposition of tungsten onto carbon lamp filaments. However, the incomplete oxidation of wood leading to the formation of the soot is the oldest example for CVD. During the 1970s, CVD was employed for the fabrication of semiconductors and coatings of electronic circuits. With advancements, the CVD process was used for the ceramics, solar cells, composites and turbine blades [105]. The material deposition is achieved by utilizing the source of energy such as light, heat and plasma. The sources of gas include chloride, bromide and hydride. The desired chemical gradient for FGM can be achieved deposition temperature, gas ratios, flow rate, gas type and gas pressure. Some of the main advantages of CVD are continuous and controlled change in composition and near net designated shape.

Functionally graded SiC/C composites were fabricated using the CVD technique by Sasaki and Hirai [106]. The fabricated FGM system was analysed for the thermal shock resistance. No cracks were observed for the FGM specimen under cyclic heat flow conditions with a surface temperatures ranging from 1,700 to 1,150 K.

Chemical vapour infiltration (CVI) technique for fabrication of FGM has the advantage of precise control of the desired chemical composition gradient and the thickness of the layers in FGM. The C/SiC FGMs were fabricated using both the CVD and CVI techniques by Kawase et al. [107]. No cracks were observed for the FGM specimens undergoing a rapid quenching test. CVI technique was used for the fabrication of carbide-cobalt-diamond FGM systems by Jain et al. [108]. Application of the electrochemical cobalt etching was reported to be the most difficult step in the fabrication of FGM system for achieving the desired pore size gradient. The pores were filled using carbon generated from the catalytic decomposition of the methane. Further, the catalytic decomposition was also used to transform carbon to diamond. Excellent toughness and wear resistance were achieved with the high amount of diamond phase on the surface.

The properties of FGMs fabricated using CVD and CVI techniques are affected by changes in the gas flow rate and the gas content. Besides the effect on properties, the thickness, microstructure and the chemical composition too are affected by the gas content and its flow rate. The effect of changing contents for CH_4 and H_2 was investigated for the fabrication of TiC/SiC/graphite FGM system by Jung et al. [109]. It was observed that the changing H_2 content could easily control the TiC, SiC and C contents in comparison to the changing CH_4 content. ZrO functionally graded systems were fabricated by Wang et al. [110] and were examined for the effect of change in the ratio of the source reaction and the rate of gas flow. One-dimensional nanowire and the two-dimensional nanoflag such as FGMs were produced. The growth of the two-dimensional functionally graded ZrO was not restricted to one direction in comparison to the one-dimensional nanowires. The physical properties of the final product were also affected by the morphology of the zinc oxide.

The process capability of CVD technique is not limited to specific material or discipline or small area of deposition. Fabrication of polycrystalline nickel films was studied by Reina et al. [111]. This was done to produce a single layer of grapheme film of dimensions ranging in centimetres. The film was found to be a suitable candidate for application in optoelectronic and electronic disciplines.

1.4.3 Thermal spray

Several patents on thermal spray were published between 1882 and 1889. A flame spray technique was used for the coatings of tin and lead by Schoop in 1911. A book on number of aspects of metal spray process and its application was published by Turner and Budgen in the year 1926, which broadened the scope of the thermal spray field. Coatings of plastic substrates and ceramics were covered in the further editions of the book [112]. A big paradigm shift in the field of thermal spray took place because of the surge in requirements for high-temperature coatings in aerospace, military and commercial sectors. The thermal spray process is being increasingly used for surface treatment of a large number of materials to produce materials with excellent wear resistance required for working tools, high thermal-resistant materials for thermal barrier applications among others [113]. A wide range of materials such as metal oxides, ceramics, metals and carbides can be used for the coatings. Several electrical sources can be used as a source of energy for the thermal spray process [114].

Different types of thermal spray process utilize different sources of energy to melt the coating materials. For instance, detonation and flame spray makes use of combustion, whereas electric current is utilized by wire arc. Furthermore, various gas temperatures are used by different processes. For example, the temperature of the gas used for plasma spray ranges from 12,000 to 16,000 °C whereas it is 3,000 °C for the powder flame spray [114].

Development of optimum processing condition for enhanced process efficiency is one of the challenging tasks in the thermal spray process. In order to establish the optimum processing conditions, experimental and numerical investigations have been conducted on different processing parameters [115, 116]. A dense and high strength structure results when the plasma spray coatings are done using low-pressure plasma spray or vacuum plasma spray.

The typical application of FGMs is that of thermal barrier coatings that are obtained using the spray process [117]. Thermal barrier coatings find applications in tungsten-coated dampers and engine components of gas turbine. However, one of the major challenges in the development of the thermal barrier coatings is that of the thermal mismatch between the base and the coating material [2, 118, 119]. Therefore, the coatings have been replaced with the FGM coatings. A marked reduction in the residual stress has been observed in case of the FGM coatings when compared to the non-graded coatings [120–122]. Impact of residual stresses on the ZrO_2/NiCrAlY FGM system has been investigated by Khor and Gu [123]. It was reported that the residual stress increases with the decreased layer thickness and the number of coatings. A finite element analysis on the same FGM system revealed that the thermal diffusivity and the thermal conductivity increased with the increasing amount of NiCrAlY [124].

Atmospheric plasma spray technique was used by Cannilo et al. [125] for the deposition of TiO_2-hydroxyapatite on the Ti_6Al_4V substrate. Increase in the Vickers hardness for the developed FGM system was reported by the optimization of the process parameters. However, a weaker interface resulted due to the recrystallization phenomenon inducing stress at higher temperatures.

Han et al. [126] made a comparison analysis for the thermal shock property of CeO_2-Y_2O_3-ZrO_2 FGM system fabricated using supersonic plasma spray and the ordinary plasma spray [126]. Enhanced thermal shock resistance was reported for the FGM system fabricated using the powder feed super plasma spray in comparison to that obtained using the ordinary spray.

Investigation on the post-treatment processes of the functionally graded thermal coatings using thermal spray was made by Stewart et al. [127]. The effect on the mechanical and the microstructural properties was studied for the hot isostatic and vacuum heating processes carried out at elevated temperatures. A higher rolling coating fatigue was reported for the thermal barrier coatings post-treated with hot isostatic. Further, a higher value of hardness and modulus was obtained as a result of better bonding strength. Shanmugavelayutham and Kobayashi [128] analysed the microstructural properties of ZrO_2-Al_2O_3 using the image analyser. The Vickers hardness of the coating showed an increasing trend with the increasing alumina content. However, crack initiation and propagation at the interface resulted, owing to the low fracture toughness of Al_2O_3.

Transmission electron microscopy (TEM), Scanning electron microscope (SEM) and Electron probe microanalyzer (EMPA) characterizations were followed for investigating the effect of plasma spray technique on generation of new source of twins and dislocations in the functionally graded MgO-ZrO/NiCrAl coatings [129]. Tendency to form oxides by Cr and Al was reported through the microstructural observations. Better mechanical stability was reported for the FGM specimen in comparison to the non-FGM specimen. Some of the other process parameters influencing mechanical, structural and physical properties of the FGMs are torch type, power, internal diameter, spraying distance, powder injection system, and plasma gas, carrier gas, cooling gas, cooling system, temperature and pressure of the substrate during deposition [125].

1.5 Conclusion

FGMs have occupied centre stage in recent times because of its wide range of applications in energy, defence, aerospace, electronics and medical fields. The present research provides a basic insight into the different fabrication methods used for the fabrication of FGM systems. Most of the research has been focused on ceramic/metal-based systems. Therefore, the research reports available on the processing techniques for the polymeric-based FGMs are limited.

There are other less utilized technique for the development of next-generation gradient structures. For instance, significant saving in time and energy for the fabrication of FGM systems is offered by the high-temperature microwave sintering process. The method offers a better thermal stress distribution control in the layers. Further, the FGMs produced using this technique have high density. The microwave sintering technique has been reported to be equally suitable for metals and ceramics as it is to the polymers.

The advancements in additive manufacturing techniques have opened new opportunities to manufacture materials with gradient compositions. The technique has the advantage of high accuracy and controllability that has allowed the fabrication of graded compositions with complex geometries. These graded structures lend themselves for applications such as biocompatible prosthesis.

References

[1] Joshi RK, Alwarappan S, Yoshimura M, Sahajwalla V, Nishina Y. Graphene oxide: the new membrane material. Appl Mater Today 2015;1(1):1–12.

[2] Kawasaki A, Watanabe R. Concept and P/M fabrication of functionally gradient materials. Ceram Int 1997;23(1):73–83.

[3] Gottron J, Harries KA, Xu Q. Creep behaviour of bamboo. Constr Build Mater 2014;66, 79–88.

[4] Shen M, Bever MB. Gradients in polymeric materials. J Mater Sci 1972;7(7):741–6.

[5] Bever MB, Duwez PE. On gradient composites. ARPA Materials Summer Conference, 1970, 117–40.

[6] Koizumi M. FGM activities in Japan. Composites Part B: Eng 1997;28(1–2):1–4.

[7] Cannillo V, Lusvarghi L, Siligardi C, Sola A. Prediction of the elastic properties profile in glass-alumina functionally graded materials. J Eur Ceram Soc 2007;27(6):2393–400.

[8] Jha DK, Kant T, Singh RK. A critical review of recent research on functionally graded plates. Compos Struct 2013;96, 833–49.

[9] Fukui Y, Takashima K, Ponton CB. Measurement of Young's modulus and internal friction of an in situ Al-Al3Ni functionally gradient material. J Mater Sci 1994;29(9):2281–8.

[10] Abbas MR, Uday MB, Noor AM, Ahmad N, Rajoo S. Microstructural evaluation of a slurry based Ni/YSZ thermal barrier coating for automotive turbocharger turbine application. Mater Des 2016;109:47–56.

[11] Dhineshkumar SR, Duraiselvam M, Natarajan S, Panwar SS, Jena T, Khan MA. Enhancement of strain tolerance of functionally graded LaTi 2 Al 9 O 19 thermal barrier coating through ultra-short pulse based laser texturing. Surf Coat Technol 2016;304:263–71.

[12] Naga SM, Awaad M, El-Maghraby HF, Hassan AM, Elhoriny M, Killinger A, Gadow R. Effect of La 2 Zr 2 O 7 coat on the hot corrosion of multi-layer thermal barrier coatings. Mater Des 2016;102:1–7.

[13] Sasaki M, Hirai T. Fabrication and properties of functionally gradient materials. J Ceram Soc Jpn 1991;99(1154):1002–13.

[14] Malik P, Kadoli R. Nonlinear bending and free vibration response of SUS_{316}–Al_2O_3 functionally graded plasma sprayed beams: theoretical and experimental study. J Vib Control 2016; DOI:1077546316659422.

[15] Merk N, Ding X, Guo X. Dc and pulse plating of CuNi and CuZn gradient foils: evaluation by SEM/TEM. Ceram Trans 1993;34 pp, 279.

[16] Lukás PA, Vrána M, Vleugels J, Anné G, Van der Biest O. Neutron diffraction studies of residual stresses in functionally graded alumina/zirconia ceramics, Mater Sci Forum 2008;309–14.

[17] Hatton B, Nicholson PS. Design and fracture of layered Al2O3/TZ3Y composites produced by electrophoretic deposition. J Am Ceram Soc 2001;84(3):571–6.

[18] Gasik SUM. Thermal-elasto-plastic analysis of W-Cu functionally graded materials subjected to a uniform heat flow by micromechanical model. J Therm Stresses 2000;23(4):395–409.

[19] Ghosh KB, Mukhopadhyay J, Basu RN. Functionally graded doped lanthanum cobalt ferrite and ceria-based composite interlayers for advancing the performance stability in solid oxide fuel cell. J Power Sources 2016;328:15–27.

[20] Müller E, Drašar Č, Schilz J, Kaysser WA. Functionally graded materials for sensor and energy applications. Mater Science Eng: A 2003;362(1):17–39.

[21] Slowak R, Hoffmann S, Liedtke R, Waser R. Functional graded high-K (Ba1– xSrx) TiO3 thin films for capacitor structures with low temperature coefficient. Integr Ferroelectr 1999;24(1–4):169–79.
[22] Birth U, Joensson M, Kieback B. Powder metallurgical processing and properties of copper/tungsten gradients. In Materials science forum. 1999;308:766–73. Trans Tech Publications.
[23] Kawasaki A, Watanabe R. Evaluation of thermomechanical performance for thermal barrier type of sintered functionally graded materials. Composites Part B: Eng 1997;28(1–2):29–35.
[24] Kawasaki A, Watanabe R. Microstructural designing and fabrication of disk shaped functionally gradient material by powder metallurgy. J Jpn Soc Powder Powder Metall 1990;37(2):253–8.
[25] Jin G, Takeuchi M, Honda S, Nishikawa T, Awaji H. Properties of multilayered mullite/Mo functionally graded materials fabricated by powder metallurgy processing. Mater Chem Phys 2005;89(2):238–43.
[26] Chenglin C, Jingchuan Z, Zhongda Y, Shidong W. Hydroxyapatite–Ti functionally graded biomaterial fabricated by powder metallurgy. Mater Science Eng: A 1999;271(1):95–100.
[27] Ma J, Tan, GEB. Processing and characterization of metal–ceramics functionally gradient materials. J Mater Process Technol 2001;113(1):446–9.
[28] Gooch WA, Chen, BHC, Burkins MS, Palicka R, Rubin JJ, Ravichandran R. Development and ballistic testing of a functionally gradient ceramic/metal applique. Mater Sci Forum 1999;308:614–21. Trans Tech Publications.
[29] Übeyli M, Balci E, Sarikan B, Öztaş MK, Camuşcu N, Yildirim RO, Keleş Ö. The ballistic performance of SiC–AA7075 functionally graded composite produced by powder metallurgy. Mater Des 2014;56:31–6.
[30] Bafekrpour E, Simon G, Naebe M, Habsuda J, Yang C, Fox B. Composition-optimized synthetic graphite/polymer nanocomposites. In ANTEC 2012, 437–41. Society of Petroleum Engineers.
[31] Bafekrpour E, Simon GP, Yang C, Habsuda J, Naebe M, Fox B. Effect of compositional gradient on thermal behavior of synthetic graphite–phenolic nanocomposites. J Therm Anal Calorim 2012;109(3):1169–76.
[32] Bafekrpour E, Simon GP, Naebe M, Habsuda J, Yang C, Fox B. Preparation and properties of composition-controlled carbon nanofiber/phenolic nanocomposites. Composites Part B: Eng 2013;52:120–6.
[33] Bafekrpour E, Simon GP, Habsuda J, Naebe M, Yang C, Fox B. Fabrication and characterization of functionally graded synthetic graphite/phenolic nanocomposites. Mater Science Eng: A 2012;545:123–31.
[34] Yuan H, Li J, Shen Q, Zhang L. In situ synthesis and sintering of ZrB_2 porous ceramics by the spark plasma sintering–reactive synthesis (SPS–RS) method. Int J Refract Met Hard Mater 2012;34:3–7.
[35] Teber A, Schoenstein F, Têtard F, Abdellaoui M, Jouini N. Effect of SPS process sintering on the microstructure and mechanical properties of nanocrystalline TiC for tools application. Int J Refract Met Hard Mater 2012;30(1):64–70.
[36] Qiao Z, Räthel J, Berger LM, Herrmann M. Investigation of binderless WC–TiC–Cr_3C_2 hard materials prepared by spark plasma sintering (SPS). Int J Refract Met Hard Mater 2013;38:7–14.
[37] Taylor G. Apparatus for making hard metal composites. USA, 1933.
[38] Cremer GD. Sintering together powders metals such as bronze, brass or aluminum. USA, 1944.
[39] Munir ZA, Anselmi-Tamburini U, Ohyanagi M. The effect of electric field and pressure on the synthesis and consolidation of materials: a review of the spark plasma sintering method. J Mater Sci 2006;41(3):763–77.
[40] Kingery WD. Sintering from prehistoric times to the present. Solid State Phenom 1992;25:1–10. Trans Tech Publications.
[41] German RM, Munir ZA. Morphology relations during bulk-transport sintering. Metall Mater Trans A 1975;6(12):2229.

[42] Sobczak JJ, Drenchev L. Metallic functionally graded materials: a specific class of advanced composites. J Mater Sci Technol 2013;29(4):297–316.
[43] Yu JH, Wang CB, Shen Q, Zhang LM. Preparation and properties of Si p/Al composites by spark plasma sintering. Mater Des 2012;41:198–202.
[44] Lee SH, Tanaka H, Kagawa Y. Spark plasma sintering and pressure less sintering of SiC using aluminum borocarbide additives. J Eur Ceram Soc 2009;29(10):2087–95.
[45] Eriksson M, Radwan M, Shen Z. Spark plasma sintering of WC, cemented carbide and functional graded materials. Int J Refract Met Hard Mater 2013;36:31–7.
[46] Hong CQ, Zhang XH Li WJ, Han JC, Meng SH. A novel functionally graded material in the ZrB_2–SiC and ZrO_2 system by spark plasma sintering. Mater Science Eng: A 2008;498(1):437–41.
[47] Mishina H, Inumaru Y, Kaitoku K. Fabrication of ZrO_2/AISI316L functionally graded materials for joint prostheses. Mater Science Eng: A 2008;475(1):141–7.
[48] Pellizzari M, Fedrizzi A, Zadra M. Influence of processing parameters and particle size on the properties of hot work and high speed tool steels by spark plasma sintering. Mater Des 2011;32(4):1796–805.
[49] Yongming L, Wei P, Shuqin L, Ruigang W, Jianqiang L. A novel functionally graded material in the Ti–Si–C system. Mater Science Eng: A 2003;345(1):99–105.
[50] Afzal MA, Kesarwani P, Reddy KM, Kalmodia S, Basu B, Balani K. Functionally graded hydroxyapatite-alumina-zirconia biocomposite: synergy of toughness and biocompatibility. Mater Science Eng: C 2012;32(5):1164–73.
[51] Zhang Z, Shen X, Zhang C, Wei S, Lee S, Wang F. A new rapid route to in-situ synthesize TiB–Ti system functionally graded materials using spark plasma sintering method. Mater Science Eng: A 2013;565:326–32.
[52] Feng H, Meng Q, Zhou Y, Jia D. Spark plasma sintering of functionally graded material in the Ti–TiB_2–B system. Mater Science Eng: A 2005;397(1):92–7.
[53] Watanabe Y, Iwasa Y, Sato H, Teramoto A, Abe K, Miura-Fujiwara E. Microstructures and mechanical properties of titanium/biodegradable-polymer FGM for bone tissue fabricated by spark plasma sintering method. J Mater Process Technol 2011;211(12):1919–26.
[54] Angerer P, Neubauer E, Yu LG, Khor KA. Texture and structure evolution of tantalum powder samples during spark-plasma-sintering (SPS) and conventional hot-pressing. Int J Refract Met Hard Mater 2007;25(4):280–5.
[55] Hulbert DM, Jiang D, Anselmi-Tamburini U, Unuvar C, Mukherjee AK. Continuous functionally graded boron carbide-aluminum nanocomposites by spark plasma sintering. Mater Science Eng: A 2008;493(1):251–5.
[56] Kambe M, Shikata H. Intensive energy density thermoelectric energy conversion system by using FGM compliant pads. Acta Astronaut 2002;51(1):161–71.
[57] El-Desouky A, Kassegne SK, Moon KS, McKittrick J, Morsi K. Rapid processing & characterization of micro-scale functionally graded porous materials. J Mater Process Technol 2013; 213(8):1251–7.
[58] Afzal MA, Kesarwani P, Reddy KM, Kalmodia S, Basu B, Balani K. Functionally graded hydroxyapatite-alumina-zirconia biocomposite: synergy of toughness and biocompatibility. Mater Science Eng: C 2012;32(5):1164–73.
[59] Jedamzik R, Neubrand A, Rödel J. Functionally graded materials by electrochemical processing and infiltration: application to tungsten/copper composites. J Mater Sci 2000;35(2):477–86.
[60] Drenchev L, Sobczak J, Sobczak N. Sedimentation phenomenon and viscosity of water–SiC suspension under gravity conditions – a water model study for composites synthesis. Colloids Surf A: Physicochemical Eng Aspects 2002;197(1):203–11.
[61] Lucignano C, Quadrini F. Indentation of functionally graded polyester composites. Measurement 2009;42(6):894–902.

[62] Koide S, Yazawa K, Asakawa N, Inoue Y. Fabrication of functionally graded bulk materials of organic polymer blends by uniaxial thermal gradient. J Mater Chem 2007;17(6): 582–90.

[63] Wilson JM, Shin YC. Microstructure and wear properties of laser-deposited functionally graded Inconel 690 reinforced with TiC. Surf Coat Technol 2012;207:517–22.

[64] Dinda GP, Dasgupta AK, Mazumder J. Laser aided direct metal deposition of Inconel 625 superalloy: microstructural evolution and thermal stability. Mater Science Eng: A 2009; 509(1):98–104.

[65] Shishkovsky I, Missemer F, Smurov I. Direct metal deposition of functional graded structures in Ti-Al system. Physics Procedia 2012;39:382–91.

[66] Schwartz RW, Schneller T, Waser R. Chemical solution deposition of electronic oxide films. CR Chim 2004;7(5):433–61.

[67] Besra L, Liu M. A review on fundamentals and applications of electrophoretic deposition (EPD). Progress Mater Sci 2007;52(1):1–61.

[68] Put S, Vleugels J, Anné G, Van der Biest O. Functionally graded ceramic and ceramic–metal composites shaped by electrophoretic deposition. Colloids Surf A: Physicochemical Eng Aspects 2003;222(1):223–32.

[69] Farnoush H, Mohandesi JA, Çimenoğlu H. Micro-scratch and corrosion behavior of functionally graded HA-TiO_2 nanostructured composite coatings fabricated by electrophoretic deposition. J Mech Behav Biomed Mater 2015;46:31–40.

[70] Farnoush H, Aldıç G, Çimenoğlu H. Functionally graded HA–TiO_2 nanostructured composite coating on Ti–6Al–4V substrate via electrophoretic deposition. Surf Coat Technol 2015;265:7–15.

[71] Katayama T, Sukenaga S, Saito N, Kagata H, Nakashima K. Fabrication of Al_2O_3-W functionally graded materials by slipcasting method. IOP Conference Series: Mater Science Eng 2011; 18(20):202023. IOP Publishing.

[72] Andertová J, Tláskal R, Maryška M, Havrda J. Functional gradient alumina ceramic materials – heat treatment of bodies prepared by slip casting method. J Eur Ceram Soc 2007;27(2): 1325–31.

[73] Yang J, Yu J, Huang Y. Recent developments in gelcasting of ceramics. J Eur Ceram Soc 2011; 31(14):2569–91.

[74] Tallon C, Franks GV. Recent trends in shape forming from colloidal processing: A review. J Ceram Soc Jpn 2011;119(1387):147–60.

[75] Young AC, Omatete OO, Janney MA, Menchhofer PA. Gelcasting of alumina. J Am Ceram Soc 1991;74(3):612–18.

[76] Garcia-Gancedo L, Olhero SM, Alves FJ, Ferreira, JM, Demoré, CE, Cochran S, Button TW. Application of gel-casting to the fabrication of 1–3 piezoelectric ceramic–polymer composites for high-frequency ultrasound devices. J Micromech Microeng 2012;22(12):125001.

[77] Park SH, Kang JH, Jung YG, Paik U. Fabrication and mechanical characterization of Al_2O_3/ZrO_2 layered composites with graded microstructure. AIP Conference Proceedings Feb 2008; 973(1):273–8.

[78] Arsha AG, Jayakumar E, Rajan, TPD, Antony V, Pai BC. Design and fabrication of functionally graded in-situ aluminium composites for automotive pistons. Mater Des 2015;88:1201–9.

[79] Watanabe Y, Kurahashi M, Kim IS, Miyazaki S, Kumai S, Sato A, Tanaka SI. Fabrication of fiber-reinforced functionally graded materials by a centrifugal in situ method from Al–Cu–Fe ternary alloy. Compos Part A: Appl Sci Manuf 2006;37(12):2186–93.

[80] Drenchev L, Sobczak J, Malinov S, Sha W. Numerical simulation of macrostructure formation in centrifugal casting of particle reinforced metal matrix composites. Part 1: model description. Modell Simul Mater Science Eng 2003;11(4):635.

[81] Drenchev L, Sobczak J, Malinov S, Sha W. Numerical simulation of macrostructure formation in centrifugal casting of particle reinforced metal matrix composites. Part 2: simulations and practical applications. Modell Simul Mater Sci Eng 2003;11(4):651.
[82] Gao JW, Wang CY. Modeling the solidification of functionally graded materials by centrifugal casting. Mater Science Eng: A 2000;292(2):207–15.
[83] Lajoye L, Suéry M. Proc. Int. Symp. On Advances in Cast Reinforced Metal Composites (ASM International 1988), pp. 15–20.
[84] Fu PX, Kang XH, Ma YC, Liu K, Li DZ, Li YY. Centrifugal casting of TiAl exhaust valves. Intermetallics 2008;16(2):130–8.
[85] Fukui Y. Fundamental investigation of functionally gradient material manufacturing system using centrifugal force. JSME Int J Ser. 3, Vib Control Eng Eng Ind 1991;34(1):144–8.
[86] Rajan TP, Pillai RM, Pai BC. Characterization of centrifugal cast functionally graded aluminum-silicon carbide metal matrix composites. Mater Charact 2010;61(10):923–8.
[87] Rodrıguez-Castro R, Wetherhold RC, Kelestemur MH. High-temperature thermo-mechanical behavior of functionally graded materials produced by plasma sprayed coating: experimental and modeling results, Met. Mater Int 2002;323(1–2):445–56.
[88] Watanabe Y, Inaguma Y, Sato H. Cold model for process of a Ni-aluminide/steel clad pipe by a reactive centrifugal casting method. Mater Lett 2011;65(3):467–70.
[89] Varma A, Rogachev AS, Mukasyan AS, Hwang S. Combustion synthesis of advanced materials: principles and applications. Adv Chem Eng 1998;24:79–226.
[90] Merzhanov AG. Combustion of solid flame. Moscow: Torus Press, 2007.
[91] Mahmoodian R, Hassan MA, Hamdi M, Yahya R, Rahbari RG. In situ TiC–Fe–Al_2O_3–TiAl/Ti_3 Al composite coating processing using centrifugal assisted combustion synthesis. Composites Part B: Eng 2014;59:279–84.
[92] Matson DM, Munir ZA. Combustion synthesis of intermetallic compounds using titanium, nickel and copper wires. Mater Science Eng: A 1992;153(1–2):700–5.
[93] Zhang XH, Han JC, Du SY, Wood JV. Microstructure and mechanical properties of TiC-Ni functionally graded materials by simultaneous combustion synthesis and compaction. J Mater Sci,(2000; 35(8):1925–30.
[94] Burkes DE, Moore JJ. Microstructure and kinetics of a functionally graded NiTi–TiC x composite produced by combustion synthesis. J Alloys Compd 2007;430(1):274–81.
[95] Menthe E, Rie KT, Schultze JW, Simson S. Structure and properties of plasma-nitrided stainless steel. Surf Coat Technol 1995;74:412–16.
[96] Larisch B, Brusky U, Spies HJ. Plasma nitriding of stainless steels at low temperatures. Surf Coat Technol 1999;116:205–11.
[97] Christiansen T, Somers MA. Low temperature gaseous nitriding and carburising of stainless steel. Surf Eng 2005;21(5–6):445–55.
[98] Zhecheva A, Sha W, Malinov S, Long A. Enhancing the microstructure and properties of titanium alloys through nitriding and other surface engineering methods. Surf Coat Technol 2005; 200(7):2192–207.
[99] Shon IJ, Munir ZA. Synthesis of TiC, TiC-Cu composites, and TiC-Cu functionally graded materials by electrothermal combustion. J Am Ceram Soc 1998;81(12):3243–8.
[100] Liang C, Ma W, Feng Z, Li C. Activated carbon supported bimetallic CoMo carbides synthesized by carbothermal hydrogen reduction. Carbon 2003;41(9):1833–9.
[101] Feng K, Chen H, Xiong J, Guo Z. Investigation on diffusion bonding of functionally graded WC–Co/Ni composite and stainless steel. Mater Des 2013;46:622–6.
[102] Chen H, Feng K, Xiong J, Luo J, Guo Z, Wang H. Characterization and forming process of a functionally graded WC–Co/Ni composite. Int J Refract Met Hard Mater 2012;35:306–10.

[103] Guo J, Fang ZZ, Fan P, Wang X. Kinetics of the formation of metal binder gradient in WC–Co by carbon diffusion induced liquid migration. Actamaterialia 2011;59(11):4719–31.
[104] De Lodyguine LS. Illuminant for incandescent lamps, America, 1893.
[105] Choy KL. Chemical vapour deposition of coatings. Progress Mater Sci 2003;48(2):57–170.
[106] Sasaki M, Hirai T. Thermal fatigue resistance of CVD SiC/C functionally gradient material. J Eur Ceram Soc 1994;14(3):257–60.
[107] Kawase M, Tago T, Kurosawa M, Utsumi H, Hashimoto K. Chemical vapor infiltration and deposition to produce a silicon carbide–carbon functionally gradient material. Chem Eng Sci 1999;54(15–16):3327–34.
[108] Jain M, Sadangi RK, Cannon WR, Kear BH. Processing of functionally graded WC/Co/diamond nanocomposites. Scriptamaterialia 2001;44(8):2099–103.
[109] Jung YG, Park SW, Choi SC. Effect of CH4 and H2 on CVD of SiC and TiC for possible fabrication of SiC/TiC/C FGM. Mater Lett 1997;30(5–6):339–45.
[110] Wang X, Chu X, Zhao H, Lu S, Fang F, Li J, et al. Controllable growth of functional gradient ZnO material using chemical vapor deposition. Integr Ferroelectr 2014;151(1):1–6.
[111] Reina A, Jia X, Ho J, Nezich D, Son H, Bulovic V, et al. Layer area, few-layer graphene films on arbitrary substrates by chemical vapor deposition. Nano Lett 2009;9(8), 3087–7.
[112] Turner TH, Budgen NF. Metal spraying: the origin, development, and applications of the metal-spray process of metallisation. C. Griffin, Limited: London, UK, 1926.
[113] Heimann RB. Plasma-spray coating: principles and applications. New Jersey, United States: John Wiley & Sons, 2008.
[114] Dorfman MR. Thermal spray applications. Adv Mater Process 2002;160(10):66–8.
[115] Pakseresht AH, Javadi AH, Nejati M, Shirvanimoghaddam K, Ghasali E, Teimouri R. Statistical analysis and multiobjective optimization of process parameters in plasma spraying of partially stabilized zirconia. Int J Adv Manuf Technol 2014;75(5–8):739–53.
[116] Pakseresht AH, Ghasali E, Nejati M, Shirvanimoghaddam K, Javadi AH, Teimouri R. Development empirical-intelligent relationship between plasma spray parameters and coating performance of yttria-stabilized zirconia. Int J Adv Manuf Technol 2015;76(5–8):1031–45.
[117] Choi KH, Kim HS, Park CH, Kim GH, Baik KH, Lee SH, et al. High-temperature thermo-mechanical behavior of functionally graded materials produced by plasma sprayed coating: experimental and modeling results. Met Mater Int 2016;22(5):817–24.
[118] Kim JH, Kim MC, Park CG. Evaluation of functionally graded thermal barrier coatings fabricated by detonation gun spray technique. Surf Coat Technol 2003;168(2):275–80.
[119] Miller RA, Lowell CE. Failure mechanisms of thermal barrier coatings exposed to elevated temperatures. Thin Solid Films 1982;95(3):265–73.
[120] Pan C, Chen B. Formation of the deformation twinning in austenitic stainless steel weld metal. J Mater Sci Lett 1995;14(24):1798–800.
[121] Pan E. Exact solution for functionally graded anisotropic elastic composite laminates. J Compos Mater 2003;37(21):1903–20.
[122] Hu C. Technical Handbook for Surface Treatment, Beijing, China: Beijing Industrial University Press, 1997.
[123] Khor KA, Gu YW. Effects of residual stress on the performance of plasma sprayed functionally graded ZrO_2/NiCoCrAlY coatings. Mater Science Eng: A 2000;277(1):64–76.
[124] Khor KA, Gu YW. Thermal properties of plasma-sprayed functionally graded thermal barrier coatings. Thin Solid Films 2000;372(1):104–13.
[125] Cannillo V, Lusvarghi L, Sola A. Production and characterization of plasma-sprayed TiO 2–hydroxyapatite functionally graded coatings. J Eur Ceram Soc 2008;28(11):2161–9.

[126] Han Z, Xu B, Wang H, Zhou S. A comparison of thermal shock behavior between currently plasma spray and supersonic plasma spray CeO_2–Y_2O_3–ZrO_2 graded thermal barrier coatings. Surf Coat Technol 2007;201(9):5253–6.
[127] Stewart S, Ahmed R, Itsukaichi T. Contact fatigue failure evaluation of post-treated WC–NiCrBSi functionally graded thermal spray coatings. Wear 2004;257(9):962–83.
[128] Shanmugavelayutham G, Kobayashi A. Mechanical properties and oxidation behaviour of plasma sprayed functionally graded zirconia–alumina thermal barrier coatings. Mater Chem Phys 2007;103(2):283–9.
[129] Pan C, Xu X. Microstructural characteristics in plasma sprayed functionally graded ZrO 2/NiCrAl coatings. Surf Coat Technol 2003;162(2):194–201.

Kamardeen O. Abdulrahman, Esther T. Akinlabi,
and Rasheedat M. Mahamood

2 Manufacturing of aluminium composite materials: A review

Abstract: Aluminium composite materials are becoming very popular as a result of their physical and mechanical characteristics, which are making them relevant to various applications. The addition of reinforcement materials with unique characteristics into aluminium produces aluminium composites with superior quality. Wear resistance, stiffness, strength and hardness are some of the improved properties obtained when reinforcement materials were added to the primary aluminium. This chapter presents some of the manufacturing processes of aluminium, its alloys and composites. The effects of reinforcements on aluminium composites from existing work and research direction on the fabrication of aluminium composite materials were discussed in this chapter.

Keywords: Composite materials, Functionally graded materials (FGMs), Manufacturing processes, Aluminium matrix, Powder metallurgy

2.1 Introduction

Aluminium comprises about 8% of the Earth crust, which makes it the second most plentiful element on the Earth's crust and it is never found in its pure state [1]. It is mostly available in its oxide form as micas, feldspars and clay. Aluminium is a light metal with a melting point of 658 °C, specific gravity of 2.7 and tensile strength between 90 and 150 MPa [2]. Aluminium, which was discovered in 1807, becomes widely commercialized after the invention of the Hall–Heroult production process in 1886 [3]. This history is summarized in Table 2.1.

Strengthening of aluminium is usually done by alloying using elements such as copper (Cu), magnesium (Mg), manganese (Mn), nickel (Ni), silicon (Si) and zinc (Zn) [4]. Addition of small amount of alloying elements changes soft and weak aluminium into a hard and strong metal and still retaining its light-weight property [2]. Some of the common types of aluminium alloys include duralumin with alloying elements composition of 3.5–4.5% Cu, 0.4–0.7% Mn and 0.4–0.7% Mg. Duralumin possesses maximum tensile strength and vastly employed in wrought conditions for stamping, forging, rivets, tubes, sheets and bars. Y-alloy is also known as copper-aluminium alloy, with alloying elements composition of 3.5–4.5% Cu, 1.2–1.7% Mn, 1.8–2.3% Ni, 0.6% Si, 0.6% Mg and 0.6% Fe. Addition of copper to pure aluminium leads to an alloy formation with better strength and machinability properties, which enable it to be used in casting and forging purposes employed in applications such as aircraft

https://doi.org/10.1515/9783110545104-002

Table 2.1: History of aluminium [3].

Year	Findings
1825	Danish chemist Ørsted produced aluminium by reducing chloride with potassium amalgam.
1827–1845	German scientist Wohler determined essential properties after successfully separating small globules.
1854	French scientist Saint-Claire Deville reduced aluminium chloride with sodium.
1856	Deville started industrial production at Nanterre using sodium.
1886	Hall and Heroult independently developed the fused electrolysis method for producing aluminium from alumina dissolved in cryolite.
1888	Industrial production of aluminium by the new electrolytic method was started in America.
1910	Industrial production started in Switzerland at Neuhausen and in France at Froges. By 1910, production was established in seven countries (Canada, France, Italy, Norway, Switzerland, the UK and the USA); total output was 45,000 tonnes.
1918	By 1918, the total world production hit 208,000 tonnes from production in nine countries.

engine piston and cylinder heads. Other types of aluminium alloys include magnalium, which comprises about 2–10% Mg and 1.75% Cu added to melted aluminium, and hindalium, an alloy formed from aluminium and magnesium with little amount of chromium, Al-Bi, Al-Pb, Al-Si-Pb and Al-Sn alloy.

The application of aluminium is based on three main properties, such as its low density of about 2.7, high mechanical strength that can be achieved using appropriate alloying elements and heat treatments, and its high corrosion resistance ability [5]. Other properties of aluminium include its good reflectability, high ductility, heat and electrical conductance, low working cost and high scrap value. Aluminium and its alloys have wide range of application as they are highly employed in sheet because of its suitability with different fabrication methods. This allows them to be machined in different ways such as sawing, drilling and shearing, to be easily formed by bending, drawing and stamping, to be joined using processes such as bolting, soldering, brazing, adhesive bonding and welding, and to be finished using processes such as painting, plating, coating, polishing and patterning. The sheets are used in cooking utensils, railings, roofing, automobile structures, aircraft skin, beverage and food cans. Aluminium plates are employed in tank cars, military vehicles and tanks, aircraft structural parts among others. While aluminium foils are employed in food packaging, pharmaceuticals and so on, casting, forging, rolling, extruding, spinning, bending, drawing, shearing, powder metal forming and squeezing are some of the most common operations used for primary shaping processes [2].

Processing of aluminium and its alloys plays an important role in determining their properties. This is because properties of aluminium alloys greatly depend on their constituents and microstructure like grain size distribution, average grain size, precipitate volume fraction and crystallographic orientation aspects such as the

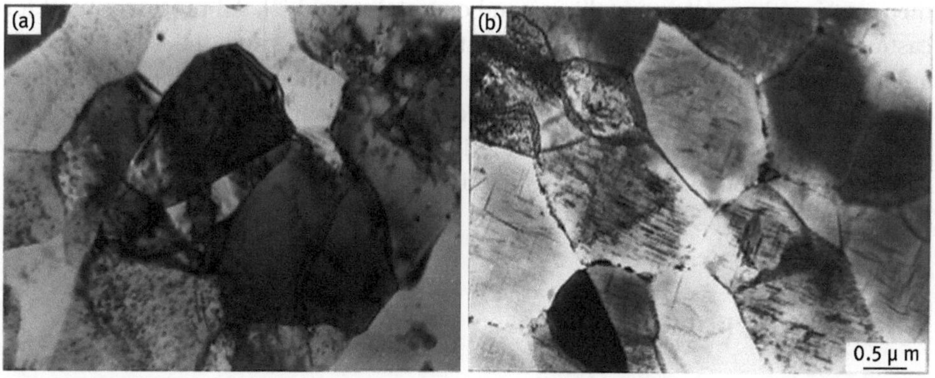

Figure 2.1: Aluminium alloy microstructure in (a) as-quenched state and (b) aged at 473 K for 20 h.

texture of aluminium and its alloys [6]. The composition is planned in a manner to cater for specific structure and texture in aluminium alloys after solidification. Such structure and texture can be achieved through the incorporation of various processing methods such as metal working fabrication methods such as forging, rolling, forming, extrusion, wire drawing, annealing and age hardening [4]. An example of microstructure of aluminium alloy with chemical composition by mass Al-2.23Cu-1.21Mg-0.93Fe-1.09Ni-0.30Sc-0.30Zr in the as-quenched state and aged at 473 K for 20 h is shown in Figure 2.1.

Apart from the fact that aluminium is one of the most widely available metals known to humans, it has proved to be very useful in areas where strength-to-weight ratio is mostly desired, especially in areas such as kitchen utensils, electronics, automobile, aircraft and military armoury. It has the ability to be successfully alloyed with other metals to give better properties and its numerous fabrication processes can be subjected to make it one of the most sought-after metal. The study is aimed at reviewing the work done on aluminium, its alloy, composites, manufacturing processes and their areas of application. Section 2.2 takes a look at various casting processes that are used for fabricating aluminium and its alloys. Other manufacturing processes for aluminium and its alloys are looked at in Section 2.3. In Section 2.4, the manufacturing processes of aluminium composite materials from various research works are reviewed. At the end, suggestions on what can further be done in this research area and conclusion are presented in Section 2.5.

2.2 Casting processes for aluminium and its alloys

Casting is one of the most important manufacturing processes that are employed in the industries, and castings are produced through remelting of ingots in furnace and then pouring the molten metal in a sand or metal mould [2]. All processes used for casting metal can be applicable for casting aluminium, and the advantages of aluminium castings lie in its ability to be produced to near-net-shape with dimensional accuracy, properties consistency and controlled surface finish [7]. Some of the processes of aluminium casting include investment casting, plaster casting, sand casting, permanent mould and pressure die casting. This section shall be looking at some of these casting processes.

2.2.1 Casting processes

Researchers are continuously interested in developing the best way possible for the casting of aluminium and its alloys using some of the well-established casting processes such as sand casting, die casting and investment casting. Sand casting involves the production of mould by ramming sand into a pattern. The removal of the pattern leaves a cavity in the sand. Sand cores are used to introduce internal cavities in casting. Poured molten metal in the mould is allowed to solidify and the mould is then fragmented to obtain the casting. Sand casting is a cheap and versatile process that makes use of different alloy ranges. Among its limitations is its relative poor surface finish and poor dimensional accuracy when compared to other manufacturing processes. However, it is flexible when it comes to the number of castings that can be produced. Figure 2.2 shows the mould for sand casting.

Permanent mould casting is another casting process that involves the pouring of molten metal through gravity into a mould of steel or cast iron. This process is similar to sand casting process. While sand moulds are removed after each casting operation, a permanent mould may be employed in achieving up to 120,000 casting cycles [8]. The process is usually employed in the casting of ferrous and non-ferrous metals. The metal moulds are much expensive to manufacture compared to moulds for sand casting.

Investment casting makes use of refractory moulds that are produced on a thermoplastic pattern or an expendable wax. The refractory slurry is enveloped

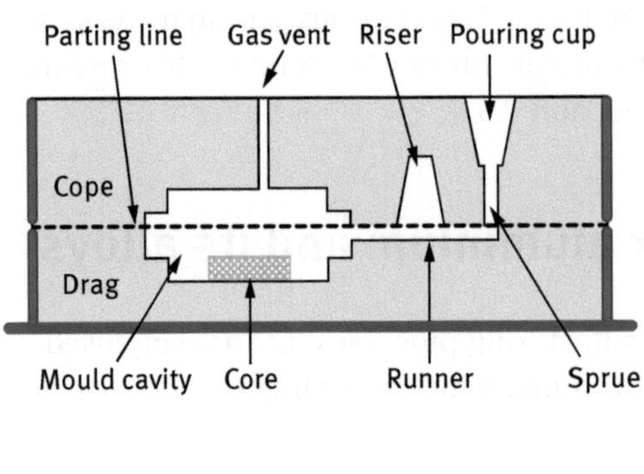

Figure 2.2: Mould for sand casting [8].

around the pattern arrangement. As the refractory dries, the pattern is melted out which then leaves behind a cavity. Molten metal is then introduced into the mould. The advantage of this process lies in the castings produced that may not require further machining. Ability of this process in producing thin walls, fine surface finish and good tolerances makes it useful in the production of precision-engineered parts and components. Figure 2.3 shows the basic steps in investment casting process.

Son et al. [10] investigated effects of neodymium (Nd) inclusion in the microstructure and the mechanical properties of Mg-5Al-3Ca alloy formed using gravity casting and extrusion process. Nd was added to the alloy to give Mg-5Al-3Ca-*x*Nd alloy (*x* from 0 to 3 mass%). The alloys were formed in a steel crucible under SF_6 and CO_2 atmosphere. The melts at a pouring temperature of 750 °C were casted into a steel mould heated to 200 °C. The as-cast alloys were then held at 380 °C for an hour and extruded into a 12 mm diameter rod at an extrusion speed of 5 mm/s with extrusion container and die temperature of 380 °C. By examination of as-cast extruded alloys under optical and scanning electron microscope with energy-dispersive X-ray spectrometer, it was observed and drawn to conclusion that as the Nd addition increases, the α-Mg matrix morphology transformed from dendritic to equiaxed grains and with average grain size reduction caused by intermetallic compounds containing Nd that was able to suppress grain growth. Also the Nd inclusion to the based alloys lead to the creation of rich Al-Nd intermetallic compounds along grain boundaries and α-Mg matrix grains, with the Al-Nd intermetallic compounds homogenously scattered. Other results obtained revealed that the hardness values of the alloy

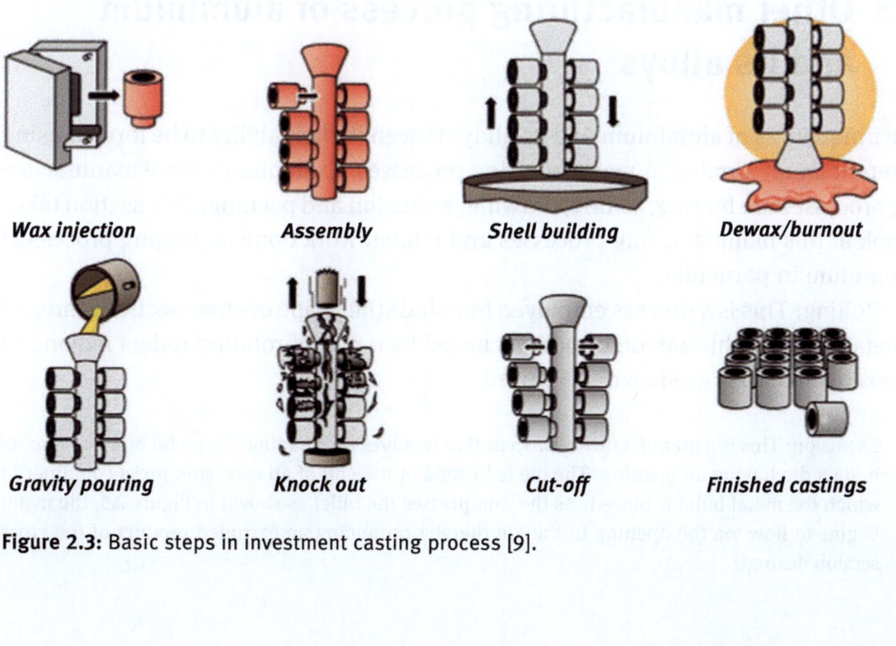

Figure 2.3: Basic steps in investment casting process [9].

were larger when compared to Mg-Al-Ca alloy without Nd. Yield strength and ultimate strength were also improved to 322 and 335 MPa, respectively.

Nakato et al. [11] developed a set of procedures in production of continuous semisolid casting of aluminium alloy billets. In this process, aluminium alloy billets (AC4C) of diameter 75 and 150 mm were cast continuously in a semisolid state, moving the alloy in an agitating container with an electromagnetic stirrer/mechanical screw. It was concluded that the solidification structure of the billets cast apart from the thin chill layer of about 2 mm thick showing a dendrite structure was majorly a mixed of fine eutectic structure and granular particles.

In the study of Raji [12], comparison of grain sizes and mechanical characteristics of Al-Si alloy components obtained through diverse casting process were analysed. Chill casting, sand casting and squeeze casting processes were utilized to produce similar shape and size of Al-8%Si alloy castings. Samples from castings were prepared and subjected to metallographic and mechanical examination. It was noticed in the microstructure that grain size of castings decreases from that of sand casting to chill casting and squeeze casting with the smallest grain size. Concurrently, mechanical characteristics of various castings improve from sand casting to chill to squeeze casting. It was, however, concluded that chill castings and squeeze castings may be utilized in as-cast condition in engineering applications requiring medium and high mechanical properties, respectively. It was then recommended that sand castings may be utilized in as cast condition in engineering applications where medium or high mechanical properties are not essential and could also be utilized in non-mechanical applications.

2.3 Other manufacturing process of aluminium and its alloys

The uniqueness of aluminium and its alloys is seen in their ability to be formed using different metal forming or manufacturing processes. Examples of these manufacturing processes are forging, rolling, drawing, extrusion and peening. This section takes a look at this manufacturing processes and related work done on forging process of aluminium in particular.

Rolling: This is a process employed to reduce the shape or cross-sectional area of a metal material through deformation caused by a pair of rotating rollers moving in opposite directions as shown in Figure 2.4.

Extrusion: This is a manufacturing process that involves the shaping of a metal billet by forcing it via a die having an opening. The die is located at the end of an extrusion press container in which the metal billet is placed. As the ram presses the billet as shown in Figure 2.5, the metal begins to flow via the opening in the die thereby producing an extruded product of the cross section desired.

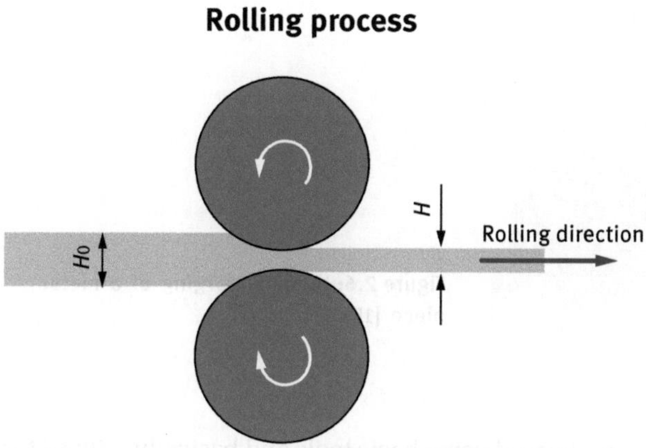

Figure 2.4: Rolling process of a metal piece [13].

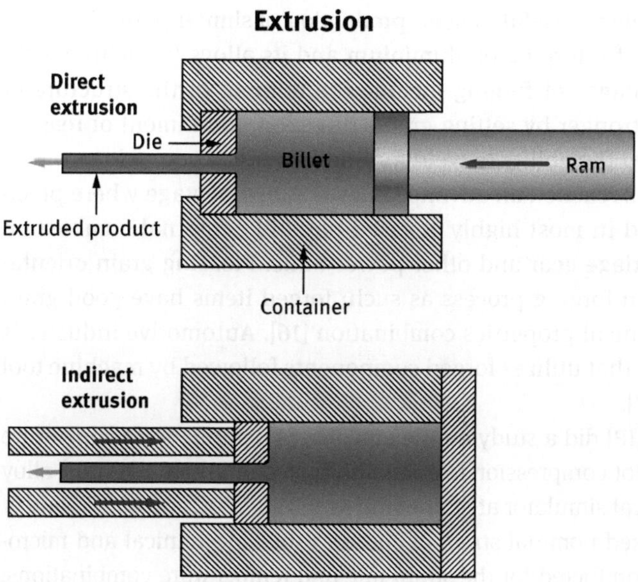

Figure 2.5: Extrusion process of a metal billet [14].

2.3.1 Forging process for aluminium and its alloys

Another manufacturing process used for aluminium and its alloys is forging. Forging involves the heating of metal to a desired temperature where sufficient plasticity is achieved and then followed by other operations such as hammering, pressing and bending to form the metal into a desired shape [2]. Some of the forging processes

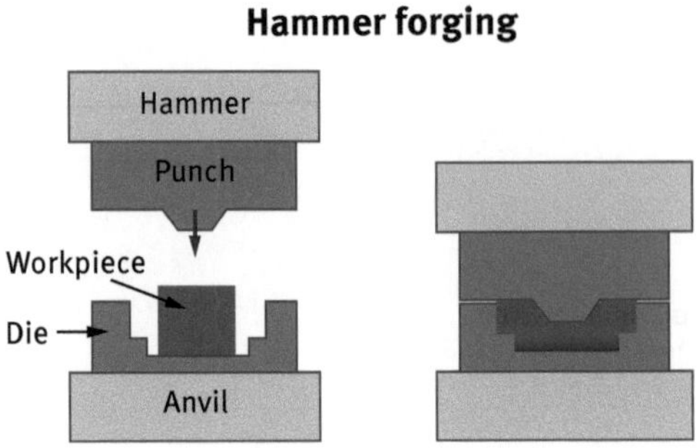

Figure 2.6: Hammer forging of a metal piece [15].

available are smith/hand forging performed using basic tools and basically utilized for little jobs, power forging that makes use of power hammers (as shown in Figure 2.6) noted for average and big jobs that need heavy blows, machine or upset forging that makes use of forging machine and the drop or stamp forging that makes use of drop hammers that make them highly useful in mass production of similar parts.

The temperature range for forging of aluminium and its alloys lies within 350–500°C. Some of the advantages of forging include the refining of the structure of metal, making the metal stronger by setting grains direction, attainment of reasonable degree of accuracy and the ability to weld forgings.

Forging technology has reached an advanced developmental stage where precision forgings are now used in most highly stressed parts like internal combustion engines, aircraft undercarriage gear and other power units. Pleasing grain orientation of metal is achieved in forging process as such; forged items have good grain structures and best mechanical properties combination [16]. Automotive industry is one of the major industries that utilizes forged components followed by machine tool and apparatus industry [17].

Dongre and Salunkhe [18] did a study on the outcome of deformation temperature on 6061 aluminium alloy. Hot compression test was conducted on Al-Mg-Si (6061) alloy on Gleeble thermomechanical simulator at temperatures of 350, 400C and 450 °C and at 0.2 and 2 strain rates for fixed nominal strain of chosen value. Mechanical and microstructural evaluation was conducted for the strain rate and temperature combinations. The true stress–strain graph indicated that at low strain rate and low temperature, flow stress is low and also at the temperature of 450 °C, low stress increases with strain rate as a result of increase of dislocation density and dislocation multiplication rate.

Vaneetveld et al. [19] conducted a study on thixoforging of aluminium alloys (7075) improvement at high solid fraction. The study utilizes a Recrystallization and partial melting (RAP) process with a seven-step heating cycles of 141 s to achieve recrystallization and improve smaller grains in the liquid matrix. Using 0.89 high solid fraction and at an extrusion ratio of 1:16, a homogeneous recrystallization was attained for slug of 35 mm diameter machined to 30 mm. High solidification rate was seen as a setback associated with the high solid fraction. The high solidification rate may be reduced by minimizing thermal exchanges between the tool and part. Some small cracks are seen due to cooling that took place in the tool after thixoforging. Raising the tool temperature to slow down the cooling parts thereby minimizing the cracks was recommended. It was concluded that to achieve an improvement in the quality of thixoforged parts, the tool temperature must be elevated in addition to a suitable punch speed to limit cracking as a result of fast cooling.

Naser et al. [20] did a study on the mechanical behaviour of multiple-forged aluminium alloy. The study was on aluminium alloy 7075 both in the multiple forged and initial states by conducting hardness, cold and hot compression test. The Vickers hardness measurements and optical microscopy images were used to evaluate the homogeneity and structure of the material. A set of constitutive expressions were obtained for the two states and the expressions were able to estimate accurately the flow stress over large range of strain rates and working temperatures. Deformation anisotropy effect is negligible during hot deformation but has a noticeable effect during cold deformation.

In the study of Rathi and Jakhade [21], different forging processes were identified and various forging defects were investigated. Repeatedly occurring forging defects, their causes and remedies were discussed. Possible causes of defects such as mismatch, unfilling and scale pits using fishbone diagram and their causes were conducted. It was finally concluded that the forging process provides better quality products when compared to other parts producing other processes. It is good to understand and control the process to prevent defects instead of discarding parts with defects at the final inspection stage.

Ball et al. [22] investigated forging residual stress influence on fatigue in aluminium. Design features like holes and machine pockets were introduced in designed and manufactured coupons in areas of different levels of bulk residual stress. Multiple methods and modelling utilizing finite element analysis were used in measuring the residual stresses at unstable areas in the coupons. Results obtained from fatigue crack growth (FCG) and fatigue crack initiation (FCI) tests from constant amplitude and spectrum loading were compared with that obtained from computed FCI and FCG. In conclusion, the work indicates that it is possible to achieve reasonable accuracy, and the influence of residual stress on fatigue using simulations.

2.4 Manufacturing of aluminium composite materials

Aluminium alloys and its composites continue to witness tremendous increase in their utilization, most importantly in places of automobile and aeronautic industries owing to their unique properties like low density, low weight, high strength, good wear resistant and coefficient of thermal expansion. Aluminium matrix composites

(AMCs) are range of high class engineering materials that can be employed in numerous applications. AMCs are made up of non-metallic reinforcements (such as silicon carbide [SiC], boron carbide [B_4C], silicon nitride [Si_3N_4], AlN, TiC, TiB_2 and TiO_2) introduced into aluminium matrix to give advantageous properties over the base metal (Al) alloys [23]. Reinforcements in AMCs are normally seen in various forms like continuous and discontinuous fibres, whisker or particulates usually in volume fractions of up to 70% [24].

Different fabrication techniques of composites are in existence. These techniques are normally based on the type of reinforced material (continuously or discontinuously) to be employed in the production of the composite. Vengatesh and Chandramohan [25] highlighted some of the techniques including stir casting technique, liquid metallurgy, gravity and squeeze casting technique.

The squeeze casting is a common technique used in the fabrication of composites of aluminium, where pressure between 70 and 150 MPa unidirectional pressure infiltration is employed. Components produced using this technique are free of void and possesses microstructure of small equiaxed grain size [26]. It is a fast process, producing good surface finishes and with the ability for selective reinforcement usage. Figure 2.7 shows the steps involved in the squeeze casting of discontinuously reinforced composites.

2.4.1 Stir casting method

Stirring is an easy production pathway for particle reinforced aluminium alloys [26]. Mechanical, electromagnetic or gas injection is employed in the stirring operation. Major issues associated with this technique are clustering or

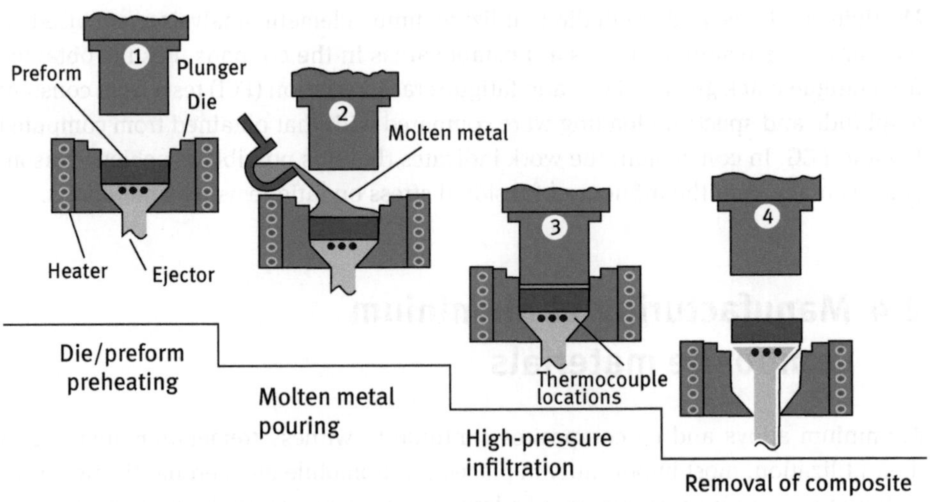

Figure 2.7: Squeeze casting of discontinuously reinforced composites [26].

agglomeration of the particles, displacement of reinforcement by liquid particles and particle redistribution that tends to affect the mechanical properties. These issues are easily taken care of employing several methods such as adding particles to the vortex created by the mixing impeller, preheating of particles before introduction, surface treatment of particles or alloying of the matrix, use of ultrasonic or electromagnetic vibration and addition of particles and metal matrix powder as pellets or briquettes. Figure 2.8 shows the schematic of a stir casting apparatus.

Meena et al. [27] analysed the mechanical properties of Al/SiC (silicon carbide) metal matrix composites (MMCs). MMC bars and circular plates fabricated using melt-stirring technique were prepared manipulating reinforced particles using 5%, 10%, 15% and 20% weight fraction. Rotating speed of the stirring process with a graphite impeller was set to 200 rev/min for 15 min. Microstructure and mechanical properties of the prepared MMCs were studied. Increase in hardness of composites was proportional to increase in reinforced particle weight fraction. The optical micrographs revealed a considerable uniform distribution of reinforced particles using the stir casting technique, leading to decrease in percentage elongation and reduction in area as the reinforced particulate size (220 mesh, 300 mesh, 400 mesh) and weight fraction increase. The impact strength also lowers with increase in the reinforced particulate size but rises with increase in weight fraction of the SiC reinforced particles.

Sambathkumar et al. [28] studied the micro-scale deformation behaviour of aluminium 7075 hybrid MMC using their two-dimensional microstructure with the help of finite element technique. Aluminium 7075 alloy was reinforced with SiC and titanium carbide (TiC) using two-step stir casting technique. Microstructure image obtained from inverted metallurgical microscope was changed to a CAD file format.

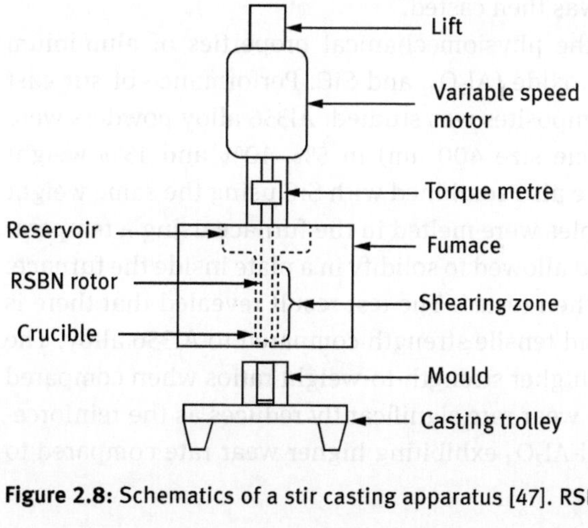

Figure 2.8: Schematics of a stir casting apparatus [47]. RSBN: Reaction bonded silicon nitride.

ABACUS 6.10, a conventional finite element tool, was then used to analyse the model. The stress–strain behaviour of the composites was analysed by varying the percentage volume of reinforced particles in ranges of 5%, 10% and 15%. The result revealed that 15% by volume percentage (SiC and TiC) aluminium composites has a maximum strength of 1,713 MPa. It was concluded that as the volume percentage rises, the strength of the hybrid MMC also rises and that 15 vol% microstructure was stronger during uniaxial tensile loading as compared to the other two.

Toptan et al. [29] carried out a study to investigate the influence of titanium inclusion on the properties of Al–B_4C interface. Average particle size of 52 µm B_4C particles utilized as reinforcement in pure aluminium (AA 1070) was used as matrix material with the addition of K_2TiF_6 flux to form a reaction layer containing TiC and TiB_2 at the interface as a way of increasing interfacial and wettability bonding. Aluminium was melted in boron nitride-coated graphite crucible to produce particulate reinforced AMCs. The microstructure of AMC specimens was studied under optical microscope and scanning electron microscope. It was discovered that due to poor wetting of B_4C particles by liquid aluminium, effective bonding was not achieved at lower temperatures as 850 °C. The addition of K_2TiF_6 flux creates a thin reaction layer of TiC and TiB_2 that solved the wetting issue. Superior mechanical and tribological properties were exhibited by Al6061-SiC composites.

Sekar et al. [30] designed a stir casting machine. The component parts such as the stirrer shaft, die, nanoparticle preheater and heating vessel were calculated for and drawn using a three-dimensional solid model. The design aids the pouring of molten metal into the die at a constant temperature before crystal growth starts because the die and the furnace are connected via a pathway taper pipe channel with heater. The machine had been used in the production of A356/Al_2O_3 nanoparticle composite materials. Sahoo et al. [31] also worked on stir casting furnace design and fabrication. The two main parts of the stir casting furnace (furnace elements and control panel) help execute the casting operation. The machine was used in the melting of aluminium scrap and the molten metal was then casted.

Sujan et al. [32] worked on the physiomechanical properties of aluminium MMCs reinforced with aluminium oxide (Al_2O_3) and SiC. Performance of stir cast Al_2O_3SiC reinforced aluminium composites was studied. Al356 alloy powders were mixed with Al_2O_3 particles (particle size 400 µm) in 5%, 10% and 15% weight fractions. Al356 alloy powders were also combined with SiC using the same weight fraction to produce Al-Al_2O_3. Samples were melted in the furnace using a temperature of 700 ° for 2 h, then stirred and allowed to solidify in a plate inside the furnace. The samples were machined and then tested. The test result revealed that there is great enhancement in hardness and tensile strength compared to Al356 alloy. The composite materials also indicate higher strength-to-weight ratios when compared with 100% aluminium. Lastly, the wear rate significantly reduces as the reinforcement particles were added with Al-Al_2O_3 exhibiting higher wear rate compared to Al-SiC composites.

Sahoo et al. [33] worked on the fabrication of aluminium alloy (Al-7Si)/titanium diboride reinforced composites using the melt stirring technique. Studies on hardness and microstructure and of direct cast composite Al-7Si-5TiB_2 and Al-7Si-5TiB_2 composite cast through cooling slope method were performed. It was discovered that the grains of TiB_2 are spheroid in shape and finer in composite obtained from the cooling slope method. The fine and spheroid nature of the grains are responsible for the increase in hardness of the composite. The study finally concluded that fabrication of Al-7Si-5TiB_2 using the stir casting technique delivered good end product, TiB_2 addition to A356 alloy increases the wear characteristic of composite and hardness.

2.4.2 Powder metallurgy

Powder metallurgy is regarded as the most common solid-state route but tends to be more expensive compared to liquid-based route [26]. Advantage of the solid-state route is the ability to achieve good mechanical properties as a result of low processing temperature that leads to low reinforcement and matrix interaction. Solidification defects such as porosity, shrinkage and segregation do away with and more consistent reinforcement distribution is achievable. Figure 2.9 shows the powder metallurgy processing route.

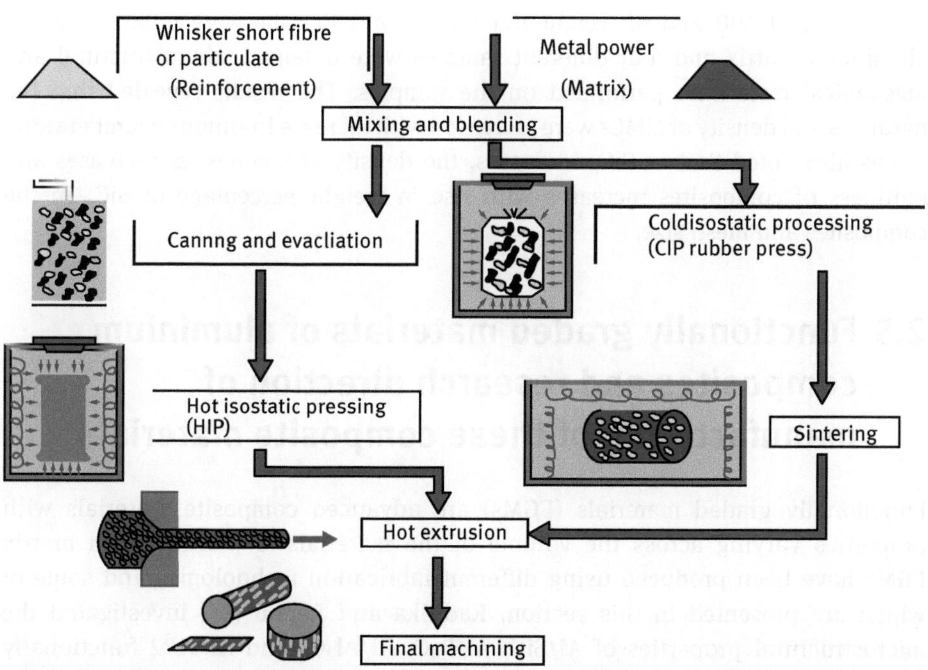

Figure 2.9: Steps for powder metallurgy processing route [26].

Chen et al. [34] carried out a study on aluminium powder size and microstructure effect on the properties of boron nitride reinforced AMCs fabricated through semi-solid powder metallurgy technique. Hexagonal boron nitride was reinforced with an aluminium alloy to form aluminium composites using semisolid powder metallurgy route. Analysis of the powder morphology and structural attribute of composites was carried out using x-ray diffraction, scanning and transmission electron microscope. It was observed that composites processed using aluminium powder with different granularity shows different grain attributes. It was also noted that there is an increase in the fracture strain and compressive strength of composites with reduction in aluminium powder size and the Brinell hardness.

Kumar et al. [35] presented experimental results on Al6061-SiC and Al7075-Al_2O_3 MMCs. Composites were produced using liquid metallurgy, where 2–6 wt% of particulates were added to the base matrix. Al6061-SiC and Al7075-Al_2O_3 composites and castings of base alloys obtained were finished by machining and subjected to tests. It was observed that the increased reinforcement's percentage increases the density and density of the composites. Micrographs of composites showed consistent distribution of particles in the samples. The tensile strength of the composites was enhanced through the dispersion of SiC and Al_2O_3 in Al6061 and Al7075 alloys, respectively.

Venkatesh and Harish [36] did a study on the mechanical properties of Al/SiC_p MMC particles fabricated through powder metallurgy route. SiC particles of mesh sizes 300 and 400 and of weight fraction 10 and 15 wt% were mixed with the aluminium matrix and four different samples were obtained. Microstructural and mechanical tests were performed on the samples. The results revealed that the hardness and density of MMCs were increased with increase in sintering temperature. It was also noted that as SiC_p increases, the density of composites increases and hardness of composites increases with rise in weight percentage of SiC_p in the composites and mesh size.

2.5 Functionally graded materials of aluminium composites and research direction of manufacturing of these composite materials

Functionally graded materials (FGMs) are advanced composite materials with properties varying across the volume of the materials [37]. Aluminium matrix FGMs have been produced using different fabrication technologies and some of which are presented in this section. Radhika and Raghu [38] investigated the microstructural properties of Al/SiC, Al/B_4C, Al/Al_2O_3 and Al/TiB2 functionally graded composites with a consistent 12% (mass fraction) reinforcement fabricated using centrifugal casting. The microstructural properties at exterior surfaces of all composites were found to contain segregation of reinforcement particles. Graded properties in terms of tensile strength and hardness of the composites were also investigated. The results showed higher hardness along the outer peripheral of all the composites with the exception in Al/B_4C composite while all the composites display high tensile strength. The wear rate was also found to be reduced in all the composites. The centrifugal casting has been found to be an important fabrication method for fabricating FGMs with a well-controlled compositional gradient taking advantage of the centrifugal force and the density differences between the materials [39]. The microstructures of the fabricated FGMs are mentioned in Figure 2.10, showing the graded microstructure of the reinforcement particles in different densities at different locations.

A similar study was carried out by Duque et al. [40], where the fabrication of aluminium alloy FGM reinforced with AlB_2 particles was done through the centrifugal casting method. The study showed that the produced FGMs have improved corrosion resistance when compared to the parent material. The study also revealed that the centrifugal casting process was effective in redistributing reinforced particles and in graded form. The process has been proved to improve the corrosion resistance of the FGM produced [41]. Powder metallurgy is another important fabrication technique employed in the fabrication of FGMs [39].

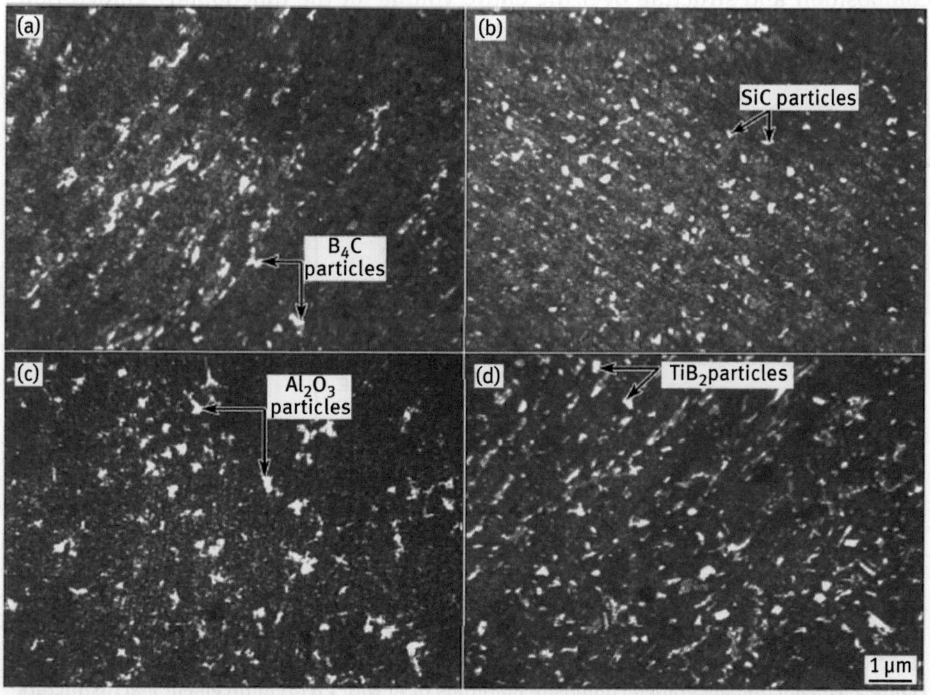

Figure 2.10: Microstructures of FGMs: (a) Al/B_4C; (b) Al/SiC; (c) Al/Al_2O_3; and (d) Al/TiB_2 [38].

Udupa et al. [42] using powder metallurgy with cold compaction method investigated the fabrication of functionally graded carbon nanotube (CNT) strengthened aluminium matrix laminates. The CNT particles were introduced to the aluminium matrix in separate weight fractions from 0.1 to 0.5 wt%. Studies on the mechanical properties of the functionally graded composites produced were carried out in order to assess how effective the fabrication method was. The results showed that the powder metallurgy technique is an effective method that can be used to alter the material properties as needed. The graded microstructure of the composite is shown in Figure 2.11, where the CNT reinforcement changes from pure Al to 0.5 wt% CNT from one end to the other end of the sample. In each layer where variation of CNT reinforcement changes from pure Al to 0.5 wt% CNT from one end to other end of the sample showed different microstructures with variation in the densities of the CNT particles. This was further confirmed by the hardness variation obtained from one end to the other with a hardness increment of up to 129% achieved with that of 0.5 wt% of CNT in the formed laminates with every layer showing strong metallurgical bonding after the sintering process with no significant microcracks or pores.

Çalıskana et al [43] carried out a similar study using the powder metallurgy technique. In this study, $Al2124/Al_2O_3$ powders of functionally graded composites were produced in various compositions. The study revealed that the transition zone can be designed to be gradual as shown in Figure 2.12 by gradually varying the composition and structure over the entire volume of the functionally graded AMC. The FGMs produced were also found to have improved properties that vary across the volume of the composites.

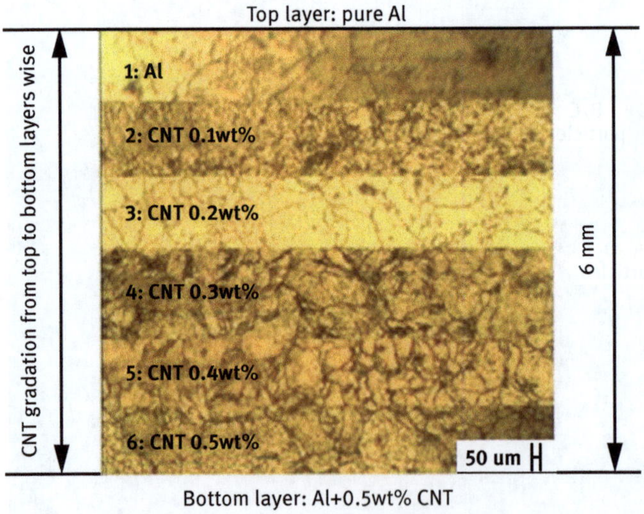

Figure 2.11: Functionally graded composite materials of Al with CNT reinforcement macrostructure with different layers [42].

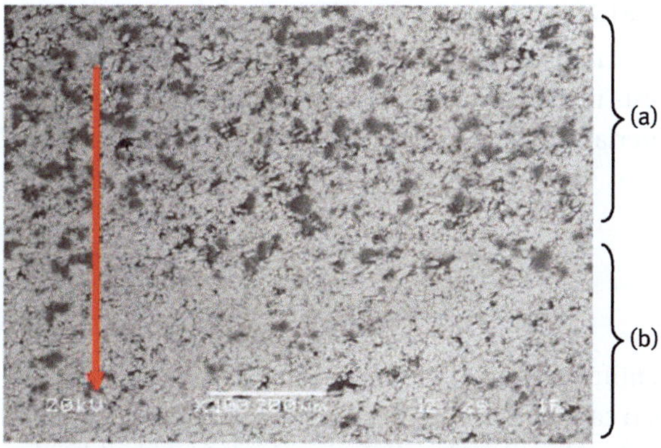

Figure 2.12: Transition zone from (a) 10 wt% Al_2O_3 reinforced material to (b) Al2124 alloy [43].

Kwon et al. [44] carried out a similar study by producing functionally graded CNT reinforced AMCs using the powder metallurgy method. The gradient layers were observed to contain various amounts of CNT as shown by diverse microstructures and hardness. The Al-CNT functionally graded composite revealed fine particle size distributions with increment in the amount of CNT addition. Hardness was found to increase with an increase in the content of CNT added. Singh and Singh [45] developed a functionally graded Al/Al_2O_3 composite using an alternative reinforced fused deposition modelling (FDM) pattern in investment casting process. The study revealed that the FGM offered better mechanical properties and improved tribological properties.

Investigation into the fabrication technique of composite materials is a vast area still experiencing continuous investigation, where possibilities on improving the properties of composites and test for most suitable means of mass production are continuously investigated. Researchers have developed fabrication techniques. These techniques cannot be taken as the only technique for composites preparation as composite fabrication is a wide area where great possibilities for creation of a better, cheaper method or new techniques that may serve as an improvement to existing ones are possible. Additive manufacturing (AM) technique such as laser metal deposition (LMD) technique is a recent area of manufacturing technique now explored in the fabrication of FGMs such as titanium and aluminium alloys [46]. The technique uses metal powder and computer-aided design model data for layer-by-layer build-up of components. Material waste reduction, reduction in downtime and production of components with unique properties are some of the advantages of this technique. Components manufactured using this technique are gaining applications in sectors such as aerospace, medicine and surgery, automobile, energy, electronics and other manufacturing industries requiring high-performance components.

The development of more powerful finite element simulation tool with great accuracy to enhance analyses of composites model is encouraged, where stress–strain behaviour by varying percentage volume of reinforced particles is achievable to study the characteristics of composites. Also development or improvement in established technique and equipment are needed in increasing interfacial and wettability bonding in composites. Finally, adequate tests including machinability are needed to be carried out on fabricated composite materials to fully establish their properties, area of application and possible failures.

2.6 Summary

Aluminium is one of the most plentiful raw materials found in the Earth's crust and never found in its pure state, as such careful practices must be done to obtain the metal in pure form. Its unique characteristics allow it to be alloyed and reinforced with other materials to produce alloys and composites with better qualities. Casting and forging are essential methods in the manufacturing of aluminium and its alloys. Production of products of near net shape with little need for subsequent machining is some of the characteristics of these manufacturing processes. The need for new products (mostly in automobile and aeronautic industry) with superior attributes such as low density, low weight, high strength and wear resistance made the production of composite materials with numerous applications an area of research that is vast. Fabrication technique of these composites depends on the type of reinforced material to be used. Notable among the techniques are stir casting, powder and liquid metallurgy and squeeze casting. AM such as LMD is a recent technique used in the fabrication of FGMs that are now finding applications in aerospace, automobile, medical, energy, electronics and other high-performance applications. Great successes are continuously recorded in the area of composites development but more still need to be done especially in carrying out an accurate and precise test in determining the properties of fabricated composites. Development of powerful tools for finite element simulations where percentage reinforcements by weight can be varied, the result is generated and studied is an interesting area that require more advancement. The fabrication methods of composites and functionally graded aluminium composite materials are presented in this chapter and the research direction has also been highlighted.

References

[1] Ashby J. The aluminium legacy: the history of the metal and its role in architecture. Constr Hist Soc 2014;15(1999):79–90.

[2] Khurmi RS, Gupta JK. A textbook of machine design, 14th ed. Ram Nagar, New Delhi: Eurasia Publishing House pvt. ltd., 2005.

[3] Budd G. Resources and production of aluminium. TALAT lecture 1101. Available at: http://core.materials.ac.uk/repository/eaa/talat/1101.pdf, 1999. Accessed date: 17th Jan. 2017.

[4] Mukhopadhyay P. Alloy designation, processing, and use of AA6XXX series aluminium alloys. ISRN Metall 2012;(Table 1:1–15, https://doi.org/10.5402/2012/165082

[5] Sheasby PG, Pinner R. The surface treatment and finishing of aluminium and its alloys, 6th edition. Materials Park, Ohio, USA: ASM International and Finishing Publications Ltd., 2001.
[6] Humphreys FJ, Hatherly M. Recrystallization and related annealing phenomena. Oxford: Elsevier, 2004.
[7] Kaufman JG, Rooy EL. Aluminum alloy castings: properties, processes, and applications. ASM Int 2004, https://doi.org/10.1017/CBO9781107415324.004.
[8] Kopeliovich D. Sand casting. Available at: http://www.substech.com/dokuwiki/doku.php?id=sand_casting, 2012. Accessed date: 15th Jan. 2017.
[9] Euroupe, Ir Cellini B. "An Introduction to the Investment Casting Process," Hitchiner Manufacturing Co., Inc. Available at: https://www.thomasnet.com/knowledge/an-introduction-to-the-investment-casting-process. Accessed date: 17th Jan. 2017.
[10] Son H-T, Lee J-S, Oh I-H, Kim D-G, Yoshimi K, Maruyama K. Microstructure and mechanical properties of Mg-Al-Ca-Nd alloys fabricated by gravity casting and extrusion process. Mater Trans 2008;49(5):1025–31, https://doi.org/10.2320/matertrans.MC200774
[11] Nakato H, Oka M, Itoyama S, Urata M, Kawasaki T, Hashiguchi K, Okano S. Continuous semi-solid casting process for aluminum alloy billets. Mater Trans 2002;43(1):24–9, https://doi.org/10.2320/matertrans.43.24
[12] Raji A. A comparative analysis of grain size and mechanical properties of AlSi alloy components produced by different casting methods. AU JT 2010;13(3):158–64. Available at: http://www.journal.au.edu/au_techno/2010/jan2010/journal133_article04.pdf
[13] Kopeliovich D. Rolling. Available at: http://www.substech.com/dokuwiki/doku.php?id=rolling, 2013c. Accessed date: 15th Jan. 2017
[14] Kopeliovich D. Extrusion. Available at: http://www.substech.com/dokuwiki/doku.php?id=extrusion, 2013a. Accessed date: 15th Jan. 2017
[15] Kopeliovich D. Forging. Available at: http://www.substech.com/dokuwiki/doku.php?id=forging, 2013b. Accessed date: 15th Jan. 2017
[16] Mahesh, NS. Forging and extrusion processes. Bangalore: Ramaiah school of advanced studies. Available at: http://164.100.133.129:81/eCONTENT/Uploads/session-8Forging processes.pdf,n.d. Accessed date: 17th Jan. 2017.
[17] Kevorkijan VM. Production technology and use of aluminum drop forging components. Mater Technol, 2001;35:191–94.
[18] Dongre RD, Salunkhe S. Study of Effect of Deformation Temperature on 6061 Aluminium Alloy by Thermo Mechanical Simulation. Glob J Res Eng A Mech Mech Eng 2014;14(2).
[19] Vaneetveld G, Rassili A, Pierret J. Improvement in Thixoforging of 7075 Aluminium Alloys at High Solid Fraction. Solid State Phenom 2008;141–43:707–12.
[20] Naser, TS. Ben, Krallics G. Mechanical behavior of multiple-forged Al 7075 aluminum alloy. Acta Polytech Hungarica 2014;11(7):103–17.
[21] Rathi MG, Jakhade NA. An Overview of forging processes with their defects. Int J Sci Res Publ 2014;4(6):1–7.
[22] Ball DL, James MA, Bucci RJ, Watton JD, Popelar CF, Bhamidipati V, et al.The impact of forging residual stress on fatigue in aluminum. Kissimmee, Florida: American Institute of Aeronautics and Astronautics, 2015.
[23] Babalola PO, Bolu CA, Inegbenebor AO, Odunfa KM. Development of aluminium matrix composites : a review. Online Int J Eng Technol Res 2014;2:1–11.
[24] Surappa MK. Aluminium matrix composites : challenges and opportunities. Sadhana 2003;28 (April):319–34.
[25] Vengatesh D, Chandramohan V. aluminium alloy metal matrix composite: survey paper. Int J Eng Res Gen Sci 2014;2(6):792–6.
[26] Froyen L, Verlinden B. Aluminium matrix composites materials. TALAT lecture 1402, 1994.

[27] Meena KL, Manna A, Banwait SS. An analysis of mechanical properties of the developed Al/SiC-MMC's. Am J Mech Eng 2013;1(1):14–19, https://doi.org/10.12691/ajme-1-1-3
[28] Sambathkumar M, Navaneethakrishnan P, Ponappa K, Sasikumar KS, Arunkumar P. Analysis of AL7075 hybrid metal matrix composite using two dimensional microstructure model based finite element method. Int J Adv Eng Technol 2016;VII(II):6–9.
[29] Toptan F, Kilicarslan A, Kerti I. The effect of Ti addition on the properties of Al-B_4C interface: a microstructural study. Mater Sci Forum 2010;636–7:192–7. https://doi.org/10.4028/www.scientific.net/MSF.636-637.192
[30] Sekar K, Allesu K, Joseph MA. Design of a stir casting machine. Am Int J Res Sci Technol Eng Math 2013;3(1):56–62.
[31] Sahoo M, Rout IS, Patra DR. Design and fabrication of a stir casting furnace set-up. Int J Eng Res Appl 2015;5(7):80–8.
[32] Sujan D, Oo Z, Rahman ME, Maleque MA, Tan CK. Physio-mechanical properties of aluminium metal matrix composites reinforced with Al_2O_3 and SiC. Int J Chem Mol Nucl Mater Metall Eng 2012;6(8):678–81.
[33] Sahoo M, Patra DR, Rout IS. Fabrication and study of titanium diboride powder and aluminium titanium alloy composite. Int J Adv Res Trends Eng Technol 2014;1(4):57–65.
[34] Chen C, Guo L, Luo J, Hao J, Guo Z, Volinsky AA. Aluminum powder size and microstructure effects on properties of boron nitride reinforced aluminum matrix composites fabricated by semi-solid powder metallurgy. Mater Sci Eng A 2015;646:306–14, https://doi.org/10.1016/j.msea.2015.08.081
[35] Kumar GB, Rao CS, Selvaraj N, Bhagyashekar MS. Studies on Al6061-SiC and Al7075-Al_2O_3 metal matrix composites. J Miner Mater Charact Eng 2010;9(1):43–55.
[36] Venkatesh B, Harish B. Mechanical properties of metal matrix composites (Al/SiCp) particles produced by powder metallurgy. Int J Eng Res Gen Sci 2015;3(1):1277–84.
[37] Mahamood RM, Akinlabi ET, Shukla M, Pityana S. Functionally graded material: an overview. Proceedings of the World Congress on Engineering (WCE 2012), III, 4–6 July 2012, London, 1593–7.
[38] Radhika N, Raghu R. Development of functionally graded aluminium composites using centrifugal casting and influence of reinforcements on mechanical and wear properties. Trans Nonferrous Met Soc China 2016;26:905–916.
[39] Mahamood RM, Akinlabi ET. Functionally graded composite materials. Springer Science Publisher, Switzerland, 2017.
[40] Duque NB, Humberto Melgarejo Z, Marcelo Suarez O. Functionally graded aluminum matrix composites produced by centrifugal casting. Mater Charact 2005,55:167–71.
[41] Velhinho A, Sequeira PD, Braz Fernándes F, Botas JD, Rocha LA. Al/SiCp functionally graded metal-matrix composites. Mater Charact 2005;55:167–71.
[42] Udupa G, Rao SS and Gangadharan KV. Fabrication of functionally graded carbon nanotube-reinforced aluminium matrix laminate by mechanical powder metallurgy technique – part I. J Mater Sci Eng 2015;4:2169–0022.
[43] Çalıskana F, Cömert S, Kocaman E. Fabrication of functional graded Al2124 composite reinforced with Al_2O_3 particles. Acta Phys Pol A 2017;131(3):437–9.
[44] Kwon H, Bradbury CR, Leparoux M. Fabrication of functionally graded carbon nanotube-reinforced aluminum matrix composite. Adv Eng Mater 2011;13(4):325–9.
[45] Singh S, Singh R. Development of functionally graded material by fused deposition modelling assisted investment casting. J Manuf Process 2016;24:38–45.
[46] Mahamood RM, Akinlabi ET, Laser metal deposition of functionally graded Ti_6Al_4V/TiC, Mater Des 2015;84:402–10.
[47] Brabazon D, Browne DJ, Carr AJ. Mechanical stir casting of aluminium alloys from the mushy state: process, microstructure and mechanical properties. Mater Sci Eng A 2002;326:370–81.

R. M. Mahamood, E. T. Akinlabi, G. M. Owolabi,
and K. O. Abdulrahman

3 Advanced manufacturing of compositionally graded composite materials: An overview

Abstract: Compositionally graded composite materials are advanced materials that contain inhomogeneous composition as well as structure that varies across the entire volume and with changing properties. Functionally graded materials are produced through different manufacturing processes, which include centrifugal method, powder metallurgy method, and additive manufacturing technology. Additive manufacturing technology is an advanced manufacturing process used for the production of three-dimensional objects simply by adding materials in layers using the digital image of the component. In this chapter, the production of compositionally graded composite materials using an advanced manufacturing method, laser additive manufacturing technologies, which include selective laser melting/sintering and laser metal deposition/laser material deposition processes are reviewed and the recent research progress is also presented.

Keywords: additive manufacturing, functionally graded materials, laser material deposition process, selective laser melting, selective laser sintering

3.1 Introduction

Compositionally graded composite materials are made up of materials with varying elemental properties as a result of varying quantities of material mixture or changing microstructural characteristic across the depth of the bulk material. The idea of functionally graded material (FGM) which presents the compositional and structural gradients in the material microstructure was initially proposed by researchers in Japan in the mid-1980s [1]. These researchers were confronted with the problem of delamination of a composite material that was used in a hypersonic space plane project that requires a thermal barrier having an external temperature of about 2,000 K and an internal temperature of about 1,000 K over a thickness of less than 10 mm thick [1]. The composite materials they were using kept failing as a result of the composite materials delaminating. They noted that the failure was happening at the interface between the composite materials due to the mismatch between the properties of these materials at the interface. They concluded that if it was possible to eliminate this sharp interface, then the problem would be solved. Then the researchers developed the novel compositionally graded composite material that was called FGMs which have a graded composition thereby eradicating the sharp interface in the initial composite material.

https://doi.org/10.1515/9783110545104-003

Compositionally graded materials have different properties both in composition and in structure throughout the entire volume of the material so as to achieve desired properties and to satisfy the intended service requirements [1–3]. FGMs have provided solutions to various engineering problems that are completely different from the application it was initially meant to solve such as in wear resistance applications. Functionally graded material is being used as an interface layer for connecting two incompatible materials in order to enhance the bond strength between the two materials. It provides the opportunity to combine different material systems such as ceramics and metals that have the ability to control properties such as deformation, dynamic response, wear, and corrosion. There are different approaches used in producing bulk compositionally graded material as well as for producing thin compositionally graded surface coating. The methods include centrifugal method, chemical vapor deposition, physical vapor deposition, additive manufacturing, and powder metallurgy [4].

Additive manufacturing method belongs to an advanced manufacturing process, used to produce three-dimensional (3D) objects through the addition of successive layers of materials [5]. Some additive manufacturing technologies are able to handle multiple materials simultaneously and hence they are able to produce 3D part that is made of compositionally graded composite materials in one manufacturing method. The most important benefit of additive manufacturing process for producing compositionally graded materials is that the part can be made directly with the required graded material composition in one manufacturing run [3]. This is not achievable in the past with the traditional manufacturing method, which involves first producing the FGMs and then makes the part by cutting the bulk material into the desired shape. The additive manufacturing process can make each layer during the building process to contain the intended desired properties along the cross section of the entire object formed or created. Additive manufacturing is able to produce the 3D object by using the 3D computer-aided design (CAD) representing the part to be produced, accepting CAD data and use it to form the object through series of materials addition in layers, following the description of the CAD data until the part building is complete [6]. This helps to build customized and personalized part with FGMs without any need for special tools that was not possible with the traditional manufacturing process. Additive manufacturing technology has developed rapidly and it has seen increasing industrial applications in the last few decades. Additive manufacturing process is attractive for its ability to shorten cycle times and the rapid transition of product concepts into physical products. Production process that usually takes weeks or months can now be produced in hours with higher accuracy. Additive manufacturing can be used to solve problems arising from production of complex parts that are often needed to be broken down into various parts when produced using the conventional manufacturing processes. By effective process control the products made of FGMs can be produced using 3D CAD model with variation of material properties that are controlled to achieve the desired. Functionally graded materials are built up with varying properties that have inhomogeneous compositions. Some additive manufacturing processes are able to incorporate various materials that made them suitable for the production of FGMs [7, 8]. The major additive manufacturing processes that are capable of producing the FGMs are sheet lamination, directed energy deposition and powder bed fusion classes of additive manufacturing technologies [4, 9, 10].

In this chapter, review on some of the recent research efforts in the fabrication of compositionally graded materials with laser additive manufacturing technologies, namely, selective laser sintering (SLS)/melting and laser material deposition process. Some of the challenges that are overcome by using this laser additive manufacturing processes for fabrication of compositionally gradient materials are discussed. The rest of this chapter is organized as follows: some research works on manufacturing of FGMs using SLS and selective laser melting (SLM) are presented in Section 3.2. Compositionally graded materials produced with the laser material deposition are explained in Section 3.3. The summary and future research direction of producing FGM using these laser additive manufacturing technologies are presented in Section 3.4.

3.2 Production of functionally graded materials with selective laser sintering/melting

SLS belongs to the powder bed fusion class of additive manufacturing technology that is based on the recent grouping according to the committee of international standard organization on additive manufacturing (committee F42) [11]. SLS produces 3D parts produced from the 3D CAD digital model representing the part by fusing the powder spread on the build platform using energy from laser [3]. The laser is used to fuse the powder while following the path dictated by the CAD data information. At the end of each cycle, the building platform is then lowered by one layer thickness. The powder is then spread on the platform and the laser scans the required area according to the data for the CAD model. The powders are sintered together with the laser and the 3D part produced is further heat treated to increase the density of the part produced. One of the important advantages of this technology is that it does not require any support structure for building the part. The unused powder serves provide the needed support for the part as the building progresses. SLM is another technology that is similar to the SLS process, the only difference is that instead of sintering the powder in the SLS, the laser fully melts the powder in the process. Also the parts produced with SLS are porous which are useful in a number of applications such as biomedical industry. SLS and SLM are used for producing prototypes, end used parts and art products. SLS and SLM are also capable of producing compositionally graded materials because the technologies allow the use of multiple materials simultaneously.

A lot of research works have been reported in the literature on this exciting technology on FGMs since their invention, and the more recent ones are reviewed in this section. The feasibility study on the production of functionally graded tissue scaffold for medical implant application has been conducted by Sudarmadji et al. [12]. Scaffolds with different graded porosity and structural configuration were produced with polycaprolactone (PCL) using SLS process. Compression test was conducted on the samples, and the degree of porosity was mapped with the resulting properties. The study showed that the compressive stiffness, the yield strength and the porosities obtained from the scaffolds produced to be closely related to those of the cancellous bone (maxillofacial region). This study has proved that SLS is feasible for the production of compositionally graded scaffolds without causing toxicity to the non-proprietary PCL material. The result is very important for tissue engineers because designing and producing suitable scaffolds have always been a great challenge. The capability of using SLS to produce tissue scaffolds will make the designing of tissue scaffolds with suitable mechanical properties possible and porosities. Functionally graded scaffolds with graded porosity with variable stiffness throughout the section of the scaffold are very important in this field and SLS will make it achievable. Traini et al. [13] also use SLS to produce functionally graded porous titanium alloy implant. The roughness, microstructure, chemical composition, mechanical properties, and fractography were investigated from the samples produced. The results showed that SLS process can be used to fabricate functionally graded porous titanium implant with the resulting properties similar to that of the human bone. This study showed that the implant with FGMs can be produced with better elastic properties that are close to the properties of human bone that will reduce stress shielding effect, and this is obtained when there is modulus of elasticity mismatch between the human bone and the implant. This will help to extend the life of the implant. In another study, SLS was used to produce 3D structure of functionally graded polymer nanocomposite material by Chung and Das [14]. Functionally graded nanocomposites made of Nylon-11 are filled with 0–10 vol% fumed silica nanoparticles. Different processing parameters were used for producing each layer in the functionally graded structure based on the result of initial experiment performed using design of experiment. The properties of FGM structures fabricated were studied. The mechanical properties were found to vary with the graded composition. The results indicate that particulate-filled compositionally graded nanocomposites can be produced by SLS process to produce 3D FGM components made of spatially varying mechanical properties. Two different 3D parts were produced to demonstrate the suitability of using SLS process for the fabrication of 3D FGM structures; they are rotator cuff scaffold and a compliant gripper. Similar study was conducted with the SLS to fabricate functionally graded Nylon-11 composite that is filled using various volume fractions of the glass beads of up to 30 vol% [15]. The study involves experimental and theoretical modeling as well as the numerical methods of analysis. The optimal SLS processing parameters used for the FGM were obtained through theoretical modeling as well as through experiments. The results obtained from experimental study were compared with the results predicted by the numerical modeling and were found to have a close agreement. The tensile strength and compressive modulus were found to increase as the volume fraction of the glass bead was increased but the strain at the yield region and break point decreased. This study also demonstrates the feasibility of producing functionally graded structure using SLS process. Salmoria et al. [16] studied the influence that hydroxyapatite (HA) content has on the properties of high-density polyethylene (HDPE)/HA composites using SLS. The result was used to fabricate functionally graded scaffold of HDPE/HA. The result of the investigation showed that SLS is suitable to fabricate HDPE/HA functional graded scaffold of controlled microstructure as well as mechanical properties that can be used for biomedical applications, for example, bone and cartilage tissues. Leite et al. [17] produced 3D functionally graded polymer blends of polyamide 12 with HDPE using SLS with compositionally gradient material in two directions. The density, microstructure and mechanical study of the specimens produced showed that the microstructure and the mechanical properties varied as the compositions. The results showed the capability of manufacturing advanced polymeric functional gradient material parts using the SLS. A number of researches have also been reported in the literature on the capability of using SLM for the fabrication of FGMs.

Maskery et al. [18] used SLM process to produce lattice structure of aluminum alloy with graded density. They performed mechanical tests on the as-deposited samples and heat-treated samples. The results showed that the mechanical properties of the produced structure can be improved with appropriate heat treatment. SLM process has also been established to be able to manufacture FGM of CoCrMo femoral stem that is lighter and flexible than the fully dense stem [19]. Hazlehurst et al. [19] also modeled the cellular structure as continuum part and determined the compressive elastic modulus using finite element method. The result obtained showed that there is good agreement between the experimental data and the model. However, SLM has been found to be limited in terms of its inability to repeatedly manufacture femoral stems from cellular structures that incorporate strut sizes that are 0.5 mm or less. Mumtaz and Hopkinson [20] presented a method for producing functionally graded nickel alloy ceramic composite material using SLM. The FGM of Waspaloy-zirconium composite was fabricated from 100 vol% Waspaloy to 10% zirconium and 90% Waspaloy. The specimens produced were studied and found to contain an average of 0.34% porosity and without any major interfacial defects.

All these research works show that the SLS/SLM is suitable for the fabrication of FGM coating and bulk FGMs. They can also be used to fabricate the required part consisting of compositionally graded materials directly from the CAD model of the part. Fabrication of compositionally graded material using the laser material deposition process is also more documented in the literature and some of them are presented in the next section.

3.3 Manufacturing of functionally graded material using laser metal deposition process

Laser material/metal deposition (LMD) technology is an important advanced manufacturing technology based on laser cladding principle. It is the class additive manufacturing named directed energy deposition [11]. LMD can create a 3D object from the 3D CAD data by adding materials layer after layer [3]. The laser beam is used to produce a melt pool on the surface of the substrate where the powder material placed coaxially with the laser is deposited. The melt pool solidifies and forms a bead upon solidification [3]. Series of beads form a long track that is seen left on the laser path. The path generated by the CAD data is followed by the laser beam to generate the 3D object by repeating the deposition layer after layer until the building of the object is complete. LMD can handle more than one material simultaneously making it useful for building part that is made up of composite materials and FGMs [3]. Another important characteristic of the LMD process is its capability to repair high-valued engineering component as well as for product remanufacturing [21]. The literature is rich on the LMD process for manufacturing of compositionally graded material both as a surface coating and as a bulk material. Some of the recent research works on LMD process for manufacturing of compositionally graded materials are presented in this section.

Thivillon et al. [22] demonstrated the potential of using the LMD process to fabricate compositionally graded coating of Co-based Stellite 6 powder and Ni-superalloy Inconel 625 powder deposited on the substrate of S235 steel. The result revealed that LMD can be used to produce FGM coating and with properties that are comparable to wrought materials. Biomedical application of FGM coating requirement is one of the driving factors of research in this area. In the research work conducted by Tanaskovic et al. [23], compositionally graded material of HA and bioglass (BG) was produced on titanium substrate. The study showed that the introduction of BG interlayers in the FGM resulted in a significant increase in the multistructure bonding onto the substrate and also the crystallinity of the HA over coating was found to be improved. The surface morphology of the FGM was also found to consist of adjacent spherical droplets which are expected to promote cell growth and proliferation. Syed et al. [24] use the copper powder and the nickel wire to produce an FGM on H13 tool steel substrate with LMD process. By analyzing the produced sample the results showed that by varying the powder feed rates of the copper powder and the nickel wire, the compositional gradients and varying properties could be achieved without the need for multipass of laser. The mechanical properties of functionally graded alumina/titanium nanolayered thin coatings produced through LMD were studied by

Bertarelli et al. [25]. The homogeneous nanocrystalline aluminum coating was produced in order to compare it with the functionally graded coating of titanium alumina. The FGM was produced by varying the percentage of alumina from 0 to 100% on the titanium substrate. The result showed that the mechanical properties of the FGM are comparable with that of the homogeneous material at low depth microindentation. Wilson and Shin [26] investigated compositionally graded material of titanium carbide (TiC) and Inconel 690 using laser metal deposition process. The compositionally graded material was produced by placing the pure Inconel 690 in one powder feeder while a premixed Inconel 690 powder and 49 vol% TiC powder were placed in another powder feeder. The powders from the two feeders were deposited on AISI 1018 steel substrate to produce the FGM. The FGM produced was found to be fully dense. The microstructure was found to be refined and a finely dispersed crystalline phase was also observed. Different volume percentage of Inconel 690 and TiC FGMs were produced in this study and a significant improvement in the microhardness and wear resistance was seen as the percentage of TiC was increased. In a similar study conducted by Wang et al. [27], compositionally graded material of titanium alloy grade 5 (Ti6Al4V) reinforced with TiC was produced using the LMD process and characterized. The FGM was fabricated with TiC powder and Ti6Al4V wire which was deposited on Ti6Al4V substrate. The TiC particles were found to be uniformly distributed in the Ti6Al4V matrix of up to volume fractions of about 74% TiC. After 74% TiC volume fraction, unmelted TiC particles were observed. The wear resistance behavior of FGMs was seen to be improved when compared to that of the substrate. The influence of the processing parameters on the properties of FGM of stainless steel SS316L and Inconel 718 produced using LMD process was studied by Shah et al. [28]. It was found that the tensile strength was increased when the powder flow rate was increased while it was found that tensile strength was inversely proportional to the laser power. The Fe_2Nb phases were seen to be formed during the deposition process. Secondary dendritic arm spacing was also found to depend on the laser power as well as the powder mass flow rate. The NbC that was observed at higher Inconel percentage ratio provides the opportunity to selectively control the mechanical properties of the compositionally graded material. The study concluded that functionally grading of stainless steel SS316L and Inconel 718 was feasible using the LMD process. Another investigation carried out by Mahamood and Akinlabi [29] also showed that FGMs can be fabricated using laser metal deposition process. They investigated the influence of producing FGMs at constant processing parameters and at changing parameters based on the compositional ratio. Two compositionally graded materials of Ti6Al4V and TiC composite were produced using laser metal deposition process. One was produced at constant process parameter while the other was produced at optimum set of process parameters for each layer based on the compositional ratio. These optimum process parameters were obtained from the model that was patented by the authors. The study showed that the FGM produced using the varying processing parameters has improved wear resistance properties when compared to the one produced at constant processing parameters [3]. Ren et al. [30] produced FGM of Ti–6Al–4V and

Ti–6.5Al–3.5Mo–1.5Zr–0.3Si on Ti6Al4V substrate using the LMD process. They studied the physical chemical and the mechanical properties of the deposited FGM. The results revealed that there is good metallurgical bonding in gradient zone with large columnar grains seen growing epitaxially in the microstructure. The chemical composition of alloying elements was found to remain unchanged in the interior of one layer, but changes abruptly in between layers which shows that there was no chemical compositional change during the melting process. The microhardness value was found to increase as the layers were increased from the substrate to the surface. The FGM was also found to have an improved tensile property. The study concludes that LMD is suitable for producing FGMs. Muller et al. [31] carried out the process modeling for the manufacturing of FGM parts using the laser metal deposition technology The model would be able to choose an appropriate manufacturing strategy as well as to also be able to control the processing parameters in order to obtain the desired properties as well as the desired part geometry. The modeling was performed based on experimental data, and the model makes it possible to control the process in a closed-loop form. Carroll et al. [32] conducted thermodynamic studies on a functionally graded component that was built from 304L stainless steel and Inconel 625 using laser metal deposition technology. The results revealed that secondary phase particles were present in the gradient zone in small quantity. The secondary phase particles were shown by the experimental results and thermodynamic calculations by transition metal carbides. The presence of these micro-sized secondary phase particles were found to be responsible for the development of cracks seen in these zones. The study showed that laser metal deposition technology has the capability to fabricate compositionally graded materials and also highlights the possible formation of secondary phases which could cause microcracks in the FGM and they need the FGM to cater for these developments as it has the potential to alter the developed properties. In a similar study, Liu and DuPont [33] created a crack-free compositionally graded composite material of pure titanium as well as TiC composite using the laser metal deposition technology. The FGM was fabricated by changing the composition from pure titanium to about 95 vol% of TiC. The two powders were placed in a separate powder feeder and through proper process control. The result of this study showed that in comparison to homogeneous Ti/TiC composite which contains cracks, the FGM composite does not contain any form of crack. The FGM was attributed to elimination of the crack formation. It was concluded that the LMD process can be used to manufacture FGM based on the ease with which the constituent materials can be delivered and the proper process control. A similar study conducted by Balla et al. [34] also showed that FGM with desired properties can be produced using laser metal deposition technology. FGMs of yttria-stabilized zirconia coatings were manufactured using laser metal deposition process. Fine microstructure seen in the coating with higher hardness was attributed to the high thermal gradients and high cooling rates produced as a result of rapid heat loss through the substrate. The substrate acts as a heat sink that resulted in the directional grain growth to be seen during the deposition process. The segmentation cracks and columnar grains observed along the coating direction were seen to be of advantage in thermal barrier application. The cracks were believed to increase the life of coatings because of expansion and contractions that will be taken place during operation. Also, the functionally graded coatings have proved to be better than the homogeneous coatings because of the higher bonding strength between the coating and the substrate. Krishna et al. [35] were able to fabricate compositionally graded material of Co–Cr–Mo alloy on Ti–6Al–4V alloy using the laser metal deposition process. A crack-free compositionally graded coating of Co–Cr–Mo/Ti–6Al–4V composite coatings from 0 to 86% Co–Cr–Mo deposited on Ti–6Al–4V substrate was successfully produced with optimal processing parameters. The FGM coating deposited on the surface of Ti6Al4V alloy was found to increase the hardness significantly. The result of human osteoblast cells in vitro cultured test showed the biocompatibility of the coatings with better bone cell proliferation. The best wear resistance and biocompatibility were seen with FGM at 50% Co–Cr–Mo alloy. This study also concludes that the LMD process can be used to manufacture FGM. Balla et al. [36] produced FGM of Ti–TiO_2 on porous Ti substrate using laser metal deposition process. A fully dense compositionally graded of up to 50% TiO_2 ceramic was deposited on porous Ti substrate.

The application of a dual structure with one side of porous structure and the other side of fully dense and hard surface is seen in medical implant such as hip prostheses. The hard side will provide the needed low friction and better surface wettability. The higher wettability will promote the formation of chemisorbed lubricating films that can help to reduce wear rate between the implant and polyethylene liner. The functionally graded coatings were also found to be biocompatible. The probability of extending mold life in the die casting industry by producing functionally graded coating on the mold using the LMD process was pursued by Ocylok et al. [37] through experimental investigations. The study showed that functionally graded coating can be build on the mold with a smooth transition from different compositions and with no cracks and low porosity can be achieved with LMD process.

Shishkovsky et al. [38] demonstrated the capability of producing FGM consisting of Ti-Al powder (an intermetallic powder) using the LMD process. The result of this research work showed that the LMD process is capable of forming heterogeneous phase of TiAl intermetallic when implemented under argon environment. The properties were found to be improved significantly. Han et al. [39] studied the possibility of producing compositionally graded properties in gear using LMD process. The processing parameters that influence on the resulting properties were also explored. The results revealed that functionally graded properties can be achieved using the LMD process and by varying the process parameters. In a similar study by Abioye et al. [40], compositionally graded material of Ti/Ni on Ti6Al4V substrate using Ni powder with commercially pure titanium wire was fabricated using the LMD process. The formation of intermetallic NiTi was found to be increased with increase in Ni content and the microhardness was also found to be decreased with increase in NiTi content. The highest microhardness was observed in the region where the $NiTi_2$ intermetallic precipitates are seen to be more. A number of other research works on the manufacturing of compositionally graded material using the LMD technology was presented by Jiang et al. [41] and Qian et al. [42] and a host of other researchers. The importance and the capability of manufacturing compositionally graded materials of bulk and functionally graded coatings have been demonstrated in this section. Lots of research works have also been carried out on the study of properties of compositionally graded materials, which is important in the further development of FGM [43].

An important characteristic of manufacturing components made up of compositionally graded materials with the additive manufacturing technology is that this component can be produced directly from the 3D CAD data, and also achieving the needed gradation which can be achieved in a single manufacturing process. This is not possible to be achieved using the traditional manufacturing process where a number of manufacturing processes are involved. There will be a need to first make the FGM in one process and then use the stock FGM to manufacture the required structure using series of other manufacturing processes depending on the complexity of the component. The additive manufacturing process that is reviewed in this research is not only capable of manufacturing functionally graded coating on an existing component but can also manufacture component no matter the complexity with graded composition, directly from the 3D CAD data of the component and in one manufacturing process. This has a number of ripple effects on the way products are designed and manufactured which is capable of manufacturing product at much cheaper rate and faster too. These important technologies are still been developed and more research work is still required in order to expand the current application of this technology. The summary and the future research need are presented in the next section.

3.4 Future research direction

Compositionally graded material is an advanced form of composite material that found its application in a wide range of industries. FGMs are found in nature such as in human body like the teeth and bone; this can be seen as the major drive for research on FGM. It has been seen that making the medical implant as close as possible to the human tissues goes a long way to increase the life of the implant as well as removing some of the problem associated with the properties mismatch between the implanted material and the surrounding human tissues. Most of the reviews presented in this study were aimed at biomedical implants. The importance of achieving medical implant with FGMs through the additive manufacturing route provides the opportunity of making highly customized components such as implants that are patient specific in a timely and cost-effective manner. Apart from the medical field, automobile and aerospace industries are also few of the industries that are in dear need of this technology because of the inherent potentials of this technology in helping to reduce the carbon footprint by these industries. A lot of research is still needed to address the repeatability issue that surrounds this technology. This challenge is one of the reasons why this technology is not fully accepted especially in application that demands proper material properties with high precision. Additive manufacturing is seen as a manufacturing process that will revolutionize the way products are designed and produced but the technology is still regarded as a niche technology because of some of the unresolved issues about this new technology. In order for the technology to gain better acceptance not only for the manufacturing of FGMs but for other components, there is need for more research works in the areas of materials characterization and process repeatability, and in the areas of process modeling and control. Understanding the physics of this important process will further help in the development of the process model that can truly represent the process which will in turn result in the proper controller design for the process. Till this is achieved, the repeatability issue is less likely to be resolved. Also more material characterization is still needed to be done which will enable the properties resulting from this process to be adequately predicted and controlled. All these research needs will be able to instill more confidence in industries that are still skeptical about the technology and also expand the application area for the industries that have already accepted the technology. Although some of the aerospace parts are now been made using this technology, but much is needed to be achieved in terms of having confidence in the properties of parts produced; then the application area can be extended to fabrication of more critical parts. A number of research works have established the influence of processing parameters on the evolving properties of LMD materials [44–51]. By changing the processing parameters, it is possible to change the properties of materials that can be varied across the volume of the material. This method of producing compositionally graded material should be developed through research to produce FGMs.

3.5 Summary

Some recent research works have been presented in this chapter to show the main research effort in the fabrication of compositionally graded materials with the laser additive manufacturing process. The laser additive manufacturing technologies that are used to produce compositionally graded parts that were reviewed in this chapter are the LMD process and the SLS/SLM. A number of research works are seen to have established the capabilities of these technologies to produce compositionally graded materials in form of coatings as well as for 3D solid parts. The future research direction has also been presented.

References

[1] Niino M, Hirai T, Watanabe R. The functionally gradient materials. J Jpn Soc Compos Mat 1987;13:257–64.
[2] Udupa G, Rao SS, Gangadharan KV. Functionally graded composite materials: an overview. Procedia Mater Sci 2014;5:1291–9.
[3] Mahamood, MR, Akinlabi ET. Functionally graded materials. Switzerland: Springer Nature, 2017.
[4] Jedamzik R, Neubrand A, Rödel J. Production of functionally graded materials from electrochemically modified carbon performs. J Am Ceram Soc 2000;83(4):983–5.
[5] Kobryn PA, Moore EH, Semiatin SL. The effect of laser power and traverse speed on microstructure, porosity and build height in laser-deposited Ti-6Al-4V. Scr Mater 2000;43:299–305.
[6] Zhou MY, Xi JT, Yan JQ. Modeling and processing of functionally graded materials for rapid prototyping. J Mater Process Technol 2004;146:396–402.
[7] Srivastava M, Maheshwari S, Kundra TK. Virtual modelling and simulation of functionally graded material component using FDM technique. Mater Today: Proc 2015;2(4–5):3471–80.
[8] Tan X, Kok Y, Tan YJ, Descoins M, Mangelinck D, Tor SB, et al. Graded microstructure and mechanical properties of additive manufactured Ti–6Al–4V via electron beam melting. Acta Mater 2015;97:1–16.
[9] Zhou MY, Xi JT, Yan JQ. Modeling and processing of functionally graded materials for rapid prototyping. J Mater Process Technol 2004;46:396–402.
[10] Wilson JM, Shin YC. Microstructure and wear properties of laser-deposited functionally graded Inconel 690 reinforced with TiC. Surf Coat Technol 2012;207:517–22.
[11] Scott J, Gupta N, Wember C, Newsom S, Wohlers T, Caffrey T. Additive manufacturing: status and opportunities. Science and Technology Policy Institute, 2012. Available at: https://www.ida.org/stpi/occasionalpapers/papers/AM3D_33012_Final.pdf. Accessed: 11 August 2016.
[12] Sudarmadji N, Tan JY, Leong KF, Chua CK, Loh YT. Investigation of the mechanical properties and porosity relationships in selective laser-sintered polyhedral for functionally graded scaffolds. Acta Biomater 2011;7:530–7.
[13] Traini T, Mangano C, Sammons RI, Mangano F, Macchi A, Piattelli A. Direct laser metal sintering as a new approach to fabrication of an isoelastic functionally graded material for manufacture of porous titanium dental implants. Dent Mater 2008;2(4):1525–33.
[14] Chung H, Das S. Functionally graded Nylon-11/silica nanocomposites produced by selective laser sintering. Mater Sci Eng A 2008;487:251–7.
[15] Chung H, Das S. Processing and properties of glass bead particulate-filled functionally graded Nylon-11 composites produced by selective laser sintering. Mater Sci Eng A 2006;437:226–34.
[16] Salmoria GV, Fancello EA, Roesler CR, Dabbas F. Functional graded scaffold of HDPE/HA prepared by selective laser sintering: microstructure and mechanical properties. Int J Adv Manuf Technol 2013;65:1529–34.
[17] Leite JL, Salmoria GV, Paggi RA, Ahrens CH, Pouzada AS. Microstructural characterization and mechanical properties of functionally graded PA12/HDPE parts by selective laser sintering. Int J Adv Manuf Technol 2012;59:583–91.
[18] Maskery I, Aboulkhair NT, Aremu OA, Tuck CJ, Ashcroft IA, Wildman RD, et al. A mechanical property evaluation of graded density Al-Si10-Mglattice structures manufactured by selective laser melting. Mater Sci Eng A 2016;670:264–74.
[19] Hazlehurst KB, Wang CJ, Stanford, M. An investigation into the flexural characteristics of functionally graded cobalt chrome femoral stems manufactured using selective laser melting. Mater Des 2014;60:177–83.

[20] Mumtaz KA, Hopkinson, N. Laser melting functionally graded composition of Waspaloy and Zirconia powders. J Mater Sci 2007;42:7647–56.
[21] Mahamood RM, Akinlabi ET, Owolabi MG. Laser Metal Deposition Process for Product Remanufacturing. In: Gupta K. editor. Advanced Manufacturing Technologies. Springer, Switzerland 2017:267–291.
[22] Thivillon L, Bertrand PH, Laget B, Smurov, I. Potential of direct metal deposition technology for manufacturing thick functionally graded coatings and parts for reactors components. J Nucl Mater 2009;385:236–41.
[23] Tanaskovic D, Jokic B, Socol G, Popescu A, Mihailescu IN, Petrovic R, Janackovic DJ. Synthesis of functionally graded bioactive glass-apatite multistructures on Ti substrates by pulsed laser deposition. Appl Surf Sci 2007;254:1279–82.
[24] Sayed WU, Pinkerton AJ, Liu Z, Li L. Coincident wire and powder deposition by laser to form compositionally graded material. Surf Coat Technol 2007;201:7083–91.
[25] Bertarelli E, Carnelli D, Gastaldi D, Tonini D, Fonzo FD, Beghi M, et al. Nanomechanical testing of alumina–titanium functionally graded thin coatings for orthopaedic applications. Surf Coat Technol 2011;205:2838–45.
[26] Wilson JM, Shin YC. Microstructure and wear properties of laser-deposited functionally graded Inconel 690 reinforced with TiC. Surf Coat Technol 2012;207:517–22.
[27] Wang F, Mei J, Wu, X. Compositionally graded Ti6Al4V + TiC made by direct laser fabrication using powder and wire. Mater Des 2007;28:2040–6.
[28] Shah K, Ul-Haq I, Khan A, Shah SA, Khan M, Pinkerton AJ. Parametric study of development of Inconel-steel functionally graded materials by laser direct metal deposition. Mater Des 2014;54:531–8.
[29] Mahamood RM, Akinlabi ET. Laser metal deposition of functionally graded Ti6Al4V/TiC, Mater Des 2015;84:402–10.
[30] Ren HS, Liu D, Tang HB, Tian XJ, Zhu YY, Wang HM. Microstructure and mechanical properties of a graded structural material. Mater Sci Eng 2014;A611:362–9.
[31] Muller P, Mognol P, Hascoet JY. Modeling and control of a direct laser powder deposition process for functionally graded materials (FGM) parts manufacturing. J Mater Process Technol 2013;213:685–92.
[32] Carroll BE, Otis RA, Borgonia JP, Suh JO, Dillon RP, Shapiro AA, et al. Functionally graded material of 304L stainless steel and inconel 625 fabricated by directed energy deposition: characterization and thermodynamic modeling. Acta Mater 2016;108:46–54.
[33] Liu W, DuPont JN. Fabrication of functionally graded TiC/Ti composites by laser engineered net shaping. Scr Mater 2003;48:1337–42.
[34] Balla VK, Bandyopadhyay PP, Bose SS, Bandyopadhyay A. Compositionally graded yttria-stabilized zirconia coating on stainless steel using laser engineered net shaping (LENSTM). Scr Mater 2007;57:861–4.
[35] Krishna BV, Xue W, Bose S, Bandyopadhyay A. Functionally graded Co–Cr–Mo coating on Ti–6Al–4V alloy structures. Acta Biomater 2008;4:697–706.
[36] Balla VK, DeVasConCellos PD, Xue W, Bose S, Bandyopadhyay A. Fabrication of compositionally and structurally graded Ti–TiO_2 structures using laser engineered net shaping (LENS). Acta Biomater 2009;5:1831–7.
[37] Ocylok S, Weisheita A, Kelbass I. Functionally graded multi-layers by laser cladding for increased wear and corrosion protection. Phys Procedia 2010;5:359–67.
[38] Shishkovsky I, Missemer F, Smurov I. Direct metal deposition of functional graded structures in Ti-Al system. Phys Procedia 2012;39:382–91.

[39] Han SW, Ji WJ, Moon YH. Fabrication of gear having functionally graded properties by direct laser melting process. Adv Mech Eng 2014;6:1–6.
[40] Abioye TE, Farayibi PK, Kinnel P, Clare AT. Functionally graded Ni-Ti microstructures synthesised in process by direct laser metal deposition. Int J Adv Manuf Technol 2015;79:843–50.
[41] Jiang W, Nair K, Molian, P. Functionally graded mold inserts by laser-based flexible fabrication: processing modeling, structural analysis, and performance evaluation. J Mater Process Technol 2005;166:286–93.
[42] Qian T, Liu D, Tian X, Liu C, Wang H. Microstructure of TA2/TA15 graded structural material by laser additive manufacturing process. Trans Nonferrous Met Soc China 2014;24:2729–36.
[43] Rahimi GH, Arefi M, Khoshgoftar MJ. Electro elastic analysis of a pressurized thick-walled functionally graded piezoelectric cylinder using the first order shears deformation theory and energy method. Mechanika 2012;18(3):292–300.
[44] Mahamood RM, Akinlabi ET, Akinlabi SA. Laser power and scanning speed influence on the mechanical property of laser metal deposited titanium-alloy. Lasers Manuf Mater Process 2015;2(1):43–55.
[45] Mahamood MR. Laser metal deposition process of metals, alloys, and composite materials. Switzerland: Springer, 2017.
[46] Pityana S, Mahamood RM, Akinlabi ET, Shukla M. Effect of gas flow rate and powder flow rate on properties of laser metal deposited Ti6Al4V. 2013 International Multi-conference of Engineering and Computer Science (IMECS 2013), March 2013, 848–51.
[47] Mahamood RM, Akinlabi ET. Effect of laser power on surface finish during laser metal deposition process. WCECS 2014;2:965–9.
[48] Shukla M, Mahamood RM, Akinlabi ET, Pityana S. Effect of laser power and powder flow rate on properties of laser metal deposited Ti6Al4V. World Acad Sci Technol 2012;6:44–8.
[49] Mahamood RM, Akinlabi ET, Akinlabi SA. Laser Power and scanning speed influence on the Mechanical property of laser metal deposited Titanium-alloy. Lasers Manuf Mater Process 2014;2(1):43–55.
[50] Mahamood RM, Akinlabi ET. Influence of scanning speed intermetallic produced in-situ in laser metal deposited TiC/Ti6Al4V composite. Mater Tehnol 2017:51(3):473–8.
[51] Mahamood RM, Akinlabi ET. Effect of processing parameters on wear resistance property of laser material deposited titanium -alloy composite. J Optoelectron Adv Mater (JOAM) 2015; 17(9–10):1348–60.

Abiodun Bayode, Esther Titilayo Akinlabi, and Sisa Pityana

4 Fabrication of stainless steel-based FGM by laser metal deposition

Abstract: Recent advancement in materials processing has resulted in the evolution of advanced composite known as functionally graded materials (FGMs). FGMs are multilayered structures with composition and/or microstructure that vary spatially across the volume of the material. This chapter presents an overview of the concept of FGMs in terms and its history. The present and potential applications of FGMs are briefly described, as well as various classification methods. Processing of FGM using laser metal deposition (LMD) is also discussed as well as the influence of LMD process parameters on the quality of laser-processed materials. Studies that have been conducted on metal–metal FGM produced by LMD are presented. Finally, a study on the microstructure and microhardness of laser deposited compositionally graded 316L/17-4PH was also presented.

Keywords: Advance composite, Functional graded material, Laser metal deposition, Microhardness, Microstructure

4.1 Introduction

There is a continuous search for new materials that are multifunctional or have better properties compared to conventional materials for modern engineering applications. Two different approaches can be adopted in the development of new materials; one way is to synthesize a unique material that is totally different from any material available or by combining two or more existing materials with dissimilar properties to form a composite. Composites have been employed for decades to successfully solve many engineering problems [1]. However, they are susceptible to stress concentration at each material interface which typically leads to delamination or debonding. To overcome this limitation, a different category of composite called functionally graded material (FGM) was developed. FGMs exhibit gradual space variations in their composition and/or microstructure from one end of the component to the other. The gradient can vary in one or several dimensions [2]. FGMs can be generally tailored at nano- to microscale compared to traditional composite which are tailored macroscale materials. FGM concept is applicable to practically all industries because of the ability to tailor materials for a specific purpose. The wide-scale use of FGMs at the moment is hindered by factors such as production cost, limited and reliable production technique. Hence, manufacturing technology is a key area of interest of many researchers.

https://doi.org/10.1515/9783110545104-004

4.1.1 Development of FGM

Biological systems have long exploited the concept of material grading to improve their performance and material usage. This type of material is plentiful in nature and exists in the form of bone, tooth, bamboo to mention a few. The bone has a unique and complex structure; the centre of the bone is denser than the outer layer. The middle layer which has lower porosity is optimized for strength, while the spongy or a more porous outer layer contains blood vessel and tissues within the cavities. Another good example is the dental crown of the tooth; the outer shell has good tribological properties, while the inner structure is brittle and ductile. For bamboo, the volume fraction of vascular bundles increases from the centre of the structure to the edges, while the size decreases. This gradation is responsible for their enhanced mechanical properties. The human skin does more than providing structure and protection to the body. It is composed of several layers of cells and tissues that serve various other purposes.

The concept of material grading was first theorized by Bever and his associates in 1972 [3]. They published two manuscripts describing possible properties, application and processing of gradient materials. However, very little progress was made in the field until the space plan project in Japan around 1984 when the first FGM was manufactured [4]. Some Japanese scientists involved in a space project required a thermal barrier coating (TBC) capable of withstanding extreme surface temperature and temperature gradients of about 2,000 and 1,000 K, respectively, over cross section of around 10 mm. There was no homogeneous or inhomogeneous material with the required heat resistance property to withstand the temperature gradient experienced during re-entry of a spacecraft. To solve this problem, Japanese scientists designed a new type of thermal barrier by improving the surface functional of the metallic base material by applying a top layer of ceramic. Both components were gradually graded thereby eliminating the sharp interface found in laminated composite. The thermal resistance of the ceramic ensured the structure was capable of withstanding high temperatures, while the metallic material provides the tensile strength, toughness and thermal conductivity required in the structure.

Other important benefits of gradient material design include increased chances of successfully bonding materials with very different coefficient of thermal expansion (CTE), reduction in residual thermal stresses, improved bond strength by elimination of sharp interfaces between constituent materials and reduction in thermal shock cracking [2, 5–7].

4.1.2 Applications of FGMs

As stated earlier, gradient materials were originally developed to eradicate the interface in traditional composites that cause large thermal stresses in the material, which

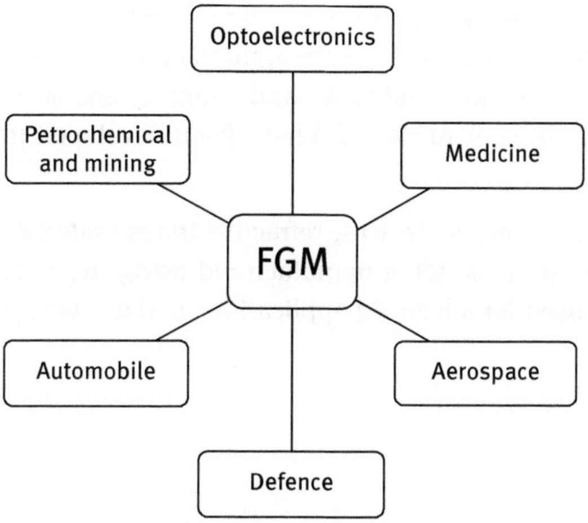

Figure 4.1: Application for FGMs.

can ultimately lead to delamination or cracking when used in high-temperature applications. Recent research has extended the FGM concept to different areas of applications in different industries and also has potential applications in the future [2, 8–10]. Figure 4.1 shows some applications of FGM.

Medicine: The use of FGMs has gained increased attention in the medical field, leading to major advances in the development of biomaterials for use in different areas of the human body. As mentioned earlier, some tissues and biological structures in the human body are functionally graded to optimize their functionality and material usage. When such materials need to be replaced because of medical emergencies, implants must be biocompatible, non-toxic and meet other functional requirements in order to serve its intended purpose. Functionally graded biomaterials can better satisfy such multiple requirements than convention materials. FGMs have been successfully used for dental implant and orthopaedic applications.

Defence: The FGM concept has also received great attention in the defence industry because of the potential of producing light-weight components which still meet the specified functional requirements. Ballistic protection for military vehicles and aircraft are areas where this concept is being applied.

aerospace: FGMs are being used in both military and commercial aircrafts to reduce thermal stress generated and also to enhance thermal resistivity. They are used as TBCs in fuselage coatings, rocket nozzles, turbine wheels and propulsion systems.

Automobile: Functionally graded ceramics are now utilized as replacements for brake disc in race car braking systems. Other areas of applications include diesel engine pistons, gas turbine engines, drive shafts, shock absorbers, power transmission and exhaust system.

Petrochemical and mining: Gradient materials show great potential in fabricating boring and cutting tools used for applications where high strength, fracture and wear resistance are required such as petroleum industry and mining industry. Functionally graded tungsten carbide and cobalt (WC/Co) have been developed to meet such a functional requirement.

Optoelectronics: Semiconductors, piezoelectric devices, refractive index materials and magnetic/optical storage devices are now being manufactured using the FGM concept. These devices are generally used for advanced applications in data storage and electronic industry.

4.1.3 Classification of FGMs

The classification of FGMs varies among researchers. FGMs have been classified based on their structure as continuous graded or stepwise graded [2]. In continuous gradation, the property of the material gradually changes from one direction to the other. While the material property changes are discontinuous in stepwise gradation. FGMs have also been categorized based on the type of gradient into three, namely porosity, microstructural and chemical composition gradient [10–12]. Porosity gradient FGMs have variable porosity across the volume of the material and typical application is in the medicine. This type of FGMs can be designed to have pores with different shapes, sizes and porosity distribution [10]. Microstructural gradation implies that FGMs have variable microstructures. This type of gradient produced in metals through surface hardening processes such as nitriding and carburation has been employed [10, 12, 13]. Chemical composition gradient can be applied to either single or multiphase materials; however, single phase materials with such gradation are rare. In multiphase material, the volume fraction of the reinforcement phase progressively varied in the bulk material. Common examples of chemical composition FGMs include ceramic–metal [14, 15] and metal–metal FGM [16, 17]. Furthermore, FGMs have been classified according to the method of fabrication. Bever and Duwez [18] groups FGM into two types: FGM produced by constructive processes and transport-based processed FGM. Similarly, FGMs have been classified according to fabrication methods as bulk and interfacial FGMs [10].

4.1.4 Manufacturing techniques

A number of manufacturing methods used in the production of FGMs are based on modifications of traditional material processing techniques. The selection of an appropriate manufacturing process depends on several factors such as materials, geometry, FGM structure (i.e. stepwise structure, continuous gradation or thin films) and the final properties of the structure. Powder metallurgy, chemical vapour deposition, centrifugal casting, thermal spraying, friction stir welding, gravity sedimentation and additive manufacturing (AM) are common methods used for manufacturing FGMs [12, 19, 20].

AM is a very attractive method for joining dissimilar materials and fabricating compositionally graded materials because fully dense parts can be created in a single step. Furthermore, AM allows for freedom in design of component and as such component with complex shapes can be manufactured relatively quicker than most FGM processes. However, AM still lags behind when compared to conventional processing methods with regard to build speed, volume, accuracy, standardization and cost. These factors have limited its adoption in the manufacturing sector for mass production of parts. At the moment, AM is used for low volume productions, especially where bespoke parts are required.

4.1.5 AM technologies

AM is a process where a component is fabricated from the bottom-up layers from a 3D computer model. AM encompasses several technologies and each of them is suitable for specific requirements and/or materials. These AM technologies have similar working principles but often differ in the way successive layers are stacked and type of heat source utilized. The available AM technologies have been categorized differently by various authors. For example, AM technologies have been classified based on the type of raw material utilized into polymer, metal and ceramic [21]. Another classification method differentiates the technologies based on nature or phase of the feedstock into liquid, powder and solid phase [21]. Most recently, various technologies and terminologies have been standardized in the ISO/ASTM 52900:2015 [22], which proposed seven distinct process types, namely vat-photo polymerization, material jetting, binder jetting, material extrusion, powder-bed fusion, sheet lamination and directed energy deposition.

Laser metal deposition (LMD) is one of the process types that falls under the direct energy deposition (DED) category. LMD has several benefits that differentiate it from the other metal AM processes such as powder bed and sheet lamination. Major benefits include multimaterial feeding and relative ease of controlling or varying material ratio [14]. Other advantages of LMD include better thermal control; minimum base metal dilution; has a relatively wide process window; can be used for repair works; easy operation in an inert environment; ability to manufacture functionally graded products; and material flexibility [23, 24]. LMD systems use a focused laser to create a melt pool on a substrate into which the metal feedstock is supplied [22]. LMD systems can be further subdivided into powder and wire-fed systems; however, hybrid systems having both wire and powder feeding have been developed [25].

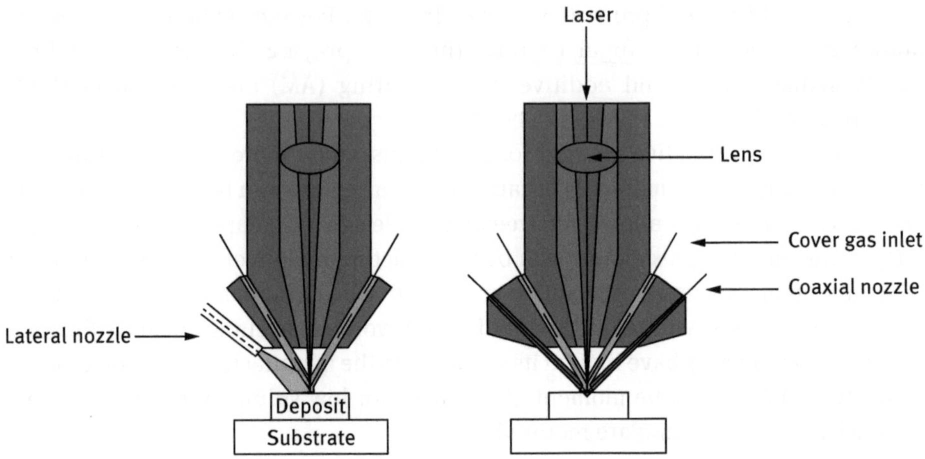

Figure 4.2: Nozzle types used in powder-fed LMD systems: (a) lateral and (b) coaxial.

The wire or powder metal can be fed using either a lateral or coaxial nozzle arrangement as shown in Figure 4.2.

Typically, powder-fed systems are more versatile as they can utilize either the lateral and coaxial nozzle arrangement. Material delivery in wire-based system is achieved mainly with lateral nozzles.

4.1.6 Process parameters

LMD is a very sensitive and complex process because the quality of the fabricated components is influenced by different processing parameters that interact with each other. The interaction between these variables must be well understood to correctly select the appropriate parameters to successfully manufacture parts with the preferred properties. There are three main groups of process parameters, namely, input parameter, process parameter and output parameter, respectively [26]. However, according to the literature, powder feed rate, laser spot size, laser power and scanning speed are the most significant parameters for laser-deposited materials. A brief description of the above-mentioned parameters are given below.

Laser spot size: This determines the beam diameter of the laser and also influences the concentration of the laser beam (large spot size tends to produce lower intensity beams and vice versa). Additionally, the width of deposited tracks is also influenced by the laser spot size.

Laser power: Laser power has a significant effect on melt pool dynamics, quality and evolving properties of a deposition. Mahamood et al. [27] presented the analysis of their study on the mechanical property and microstructure of laser-consolidated Ti6Al4V/TiC composite. The composite was processed by varying the laser power between 0.8 and 3.0 kW. Laser power was observed to influence the microhardness, grain structure and dilution of the fabricated component. Shukla et al. [28] observed a similar trend in their experiment. They found that higher laser power caused an increase in microhardness. However, Bayode et al. [29] found in their study that laser power had an inverse relationship with microhardness in laser-consolidated AISI 316L powder. The surface finish of Ti6Al4V powder processed by LMD was investigated by Mahamood and Akinlabi [30]. They observed a reduction in surface roughness as the laser power increased.

Scanning speed: This describes how fast a laser beam travels along a predetermined path. The scan speed affects the laser-material interaction time and the cooling rate during solidification. The scanning speed should neither be too fast or too slow. Depending on the laser power, shallow melt pools are formed at a very high scanning speed due to insufficient energy input, while a deeper melt pool is created at extremely low scan speeds. Akinlabi et al. [31] investigated the influence of scan speed on material efficiency. The results revealed that powder efficiency decreased as the travel speed increased. The lowest speed (0.01 m/s) produced the highest powder efficiency, while the lowest efficiency (24.82%) was obtained at 0.005 m/s scanning speed. Similarly, Mahamood et al. [32] investigated the influence of scan speed on deposition efficiency and clad height of the laser-processed Ti6Al4V alloy. All parameters were fixed except for scan speed, which was uniformly increased from 0.005 to 0.095 m/s. No clear relationship was found between scanning speed and powder efficiency. However, optimum material efficiency occurred when the speed was set at 0.045 m/s.

Powder flow rate: This refers to the amount of powder discharged from the nozzle per unit time. Powder flowability is dependent on several factors such as particle morphology and particle size distribution. Generally, powder flow rate influences dimensional accuracy and material efficiency of LMD-processed parts [28, 33, 34]. Powder flow rate has also been reported to influence porosity, microhardness and microstructure of LMD-processed Ti6Al4V/Cu composites [35]. Surface roughness of deposition tends to increase as powder flow rate increases, because the exposure time to laser irradiation of the powder particles is small at high powder flow rate but more when the flow rate is low [34].

It is noteworthy to mention that laser spot size, laser power and scan speed have a strong influence on each other and together. These variables regulate the amount of energy supplied per unit area during LMD. The relationship between these three quantities determines the energy supplied per unit area or laser energy density (LED). LED is defined by the following relationship [36]:

$$\text{Laser energy density}\left(\frac{J}{\text{mm}^2}\right) = \frac{P}{V \times D} \tag{1}$$

where P is the laser power (W), V is the scanning speed (mm/s) and D is the laser spot size (mm). LED is also a vital factor in LMD and its effects are well reported in the literature. For example, Mahamood et al. [37] in their study found that energy density and surface roughness had an inverse relationship. However, microhardness of the samples varied linearly with LED.

4.1.7 Laser-deposited FGM

There are very limited studies on metal–metal FGM in the literature when compared to metal–ceramic FGMs. Most of these studies focus on verifying the feasibility of processing FGM using LMD, identifying optimal processing conditions and also investigating their evolving properties. Wu et al. [16] investigated the microstructural evolution and mechanical properties of AISI 316/Inconel 718 FGM fabricated laser rapid manufacturing. The aim of their research was to determine the feasibility of consolidating both metal powders and also report on the evolving properties of the fabricated component. Two different types of FGMs were produced by varying their compositions and build direction. Different solidification modes were observed in both FGMs, and cellular to dendritic morphologies were observed at different regions of the FGMs. The microhardness of FGMs was also measured at different sections of the samples. An initial reduction in microhardness was observed between layers 1 and 4 of the traditional FGM; however, as the Inconel 718 content increased to about 40%, the hardness value started to increase. On the other hand, the microhardness value of FGM 2 showed an increase along the graded direction in both the longitudinal and transverse directions.

Sahasrabudhe et al. [38] investigated the microhardness and microstructure of stainless steel (SS) and titanium bimetal structure fabricated by LENS. Two different types of FGMs were fabricated in this study; one was fabricated by simply varying the compositions of 316L and Ti64 powders, while the other was almost identical to the first FGM except an intermediate layer made up of NiCr was added. The FGM produced without the NiCr layer was characterized by cracks and delaminated as the volume fraction of Ti64 increased, while the specimen with NiCr interlayer had no cracks but porosity was observed in certain areas of the structure. The authors reported that the delamination and cracks could be attributed to the formation of brittle intermetallic phases such as Cr_2Ti, Fe_2Ti and FeTi in the microstructure.

Lin et al. [17] investigated the evolving properties of compositionally graded AISI 316L/Rene88DT. The structure was characterized mostly by dendritic growth except for region of the deposit having a 100% Rene88DT super alloy content. Lin et al. [39] conducted another experiment using laser deposition to manufacture compositionally graded titanium and Rene88DTcomposite. The structure was characterized by epitaxial growth and columnar/equiaxed transition in the microstructure.

Noecker et al. [40] conducted a study on LMD of functional graded copper and steel bimetal. The aim was to enhance the thermal conductivity of steel moulds through the introduction of Cu to the component. Their attempt at compositionally grading Cu to tool steel was hindered by solidification cracking. They observed a relationship between solidification cracking and Cu concentration in the built part. They categorized cracking susceptibility into three: low, medium and high susceptibilities.

Carroll et al. [41] studied the feasibility of producing AISI 304L/Inconel 625 FGM by DED. They investigated the evolving properties of the FGM and also did thermodynamic computational modelling of the fabricated component. The results of the authors attempt to fabricate and characterize a compositionally graded 316L/17-4PH bimetal composite synthesized by LMD is presented in the next section.

4.2 Experimental

This section describes the equipment and materials used in this study to fabricate the compositionally graded bimetal and also presents the methodology used in evaluating the mechanical and microstructural properties of the laser-processed component.

4.2.1 Materials

Two different varieties of SS alloys such as 17-4PH and AISI 316L SS powders were used in this study. The 17-4PH powder is a martensitic and was manufactured by gas atomization process, while the 316L alloy is austenitic and water atomized. According to manufacturer's label, 17-4PH and 316L powders have particle sizes of 45–90 and 44 µm, respectively, while AISI 316 is an austenitic SS and non-magnetic. The SS alloy composition provided by the suppliers is presented in Table 4.1.

Table 4.1: Elemental composition of the powders

Element	17-4PH (wt%)	316L (wt%)
Iron	73.7	67.5
Nickel	4.4	13.0
Manganese	0.9	–
Chromium	16.4	17.0
Copper	4.0	–
Carbon	0.01	–
Silicon	0.7	–
Niobium	0.32	–
Molybdenum	–	2.5

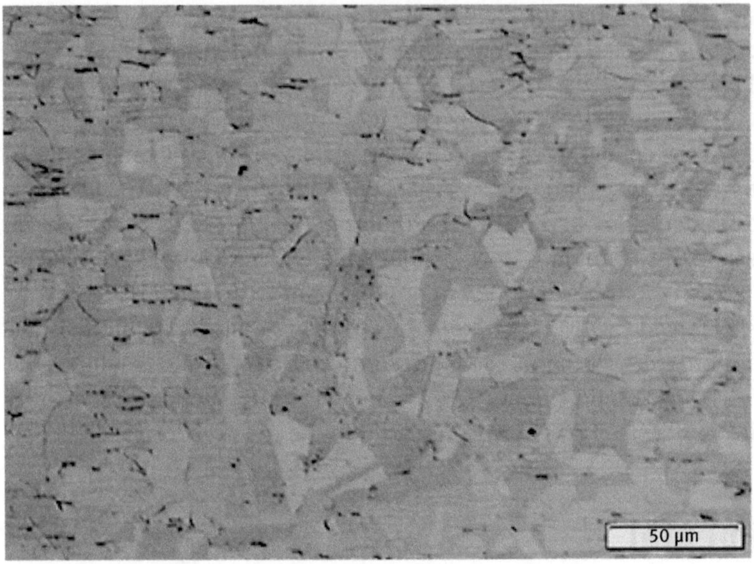

Figure 4.3: Optical micrograph of AISI 316 plate.

The substrate used was AISI 316 stainless steel plate having a dimension of 100 × 100 × 10 mm and was supplied by metal centre (Pty), South Africa. Before laser deposition, the substrate was sand blasted and then cleaned with acetone and water. This was done for two reasons, first to rid the substrate surface of any pollutant, and lastly, to increase laser absorptivity of the plate surface. The AISI 316 stainless steel substrate has purely austenitic microstructure with clearly defined grain boundary as presented in Figure 4.3.

4.2.2 Experimental set-up

The FGM was fabricated using the LMD machine at the Council for Scientific and Industrial Research laser laboratory. The LMD system consists of the following key components: laser source, powder feeder, laser head and motion device. The laser used in this study was a 3.0 kW continuous wave Nd:YAG laser operating at a wavelength of about 1.06 µm. AGTV PF 2/2 powder feeder with two hoppers that are separately driven and independently controlled was used for material delivery. A laser deposition head comprising a coaxial delivery nozzle and laser beam focusing lens assembly was mounted on a KUKA industrial robot for motion. Argon was used as both the carrier and cover gas. A picture of the LMD machine is displayed in Figure 4.4.

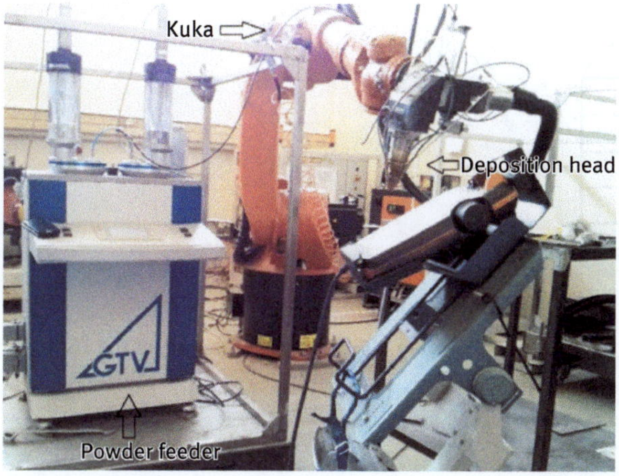

Figure 4.4: LMD system.

4.2.3 Processing

FGM was fabricated by directing the laser beam on the substrate creating a molten pool into which the stainless steel powders were supplied. The melt pool increases in size as a result of added metal and solidifies to form a solid track or deposited layer. The deposition head is moved incrementally along the *z*-axis to deposit new layers until the build is completed. The powder ratio for both stainless steel powder was varied uniformly from 100% 17-4PH to 100% 316L stainless steel alloy by adjusting the flow rate of both powders to produce a nine-layered structure as illustrated in Figure 4.3. The processing parameter used to fabricate the FGM is presented in Table 4.2. An FGM specimen with nine layers was produced as illustrated in Figure 4.5.

Table 4.2: Processing parameters.

Sample	Powder flow rate (rev/min) (17-4PH)	Powder flow rate (rev/min) (316L)	Laser power (W)	Scanning speed (m/min)	Gas flow rate (l/min)	Beam diameter (mm)	Overlap percentage (%)
A1	2.0	0	2400	0.8	2.54	2	50
	1.6	0.4					
	1.2	0.8					
	0.8	1.2					
	0.4	1.6					
	0	2.0					

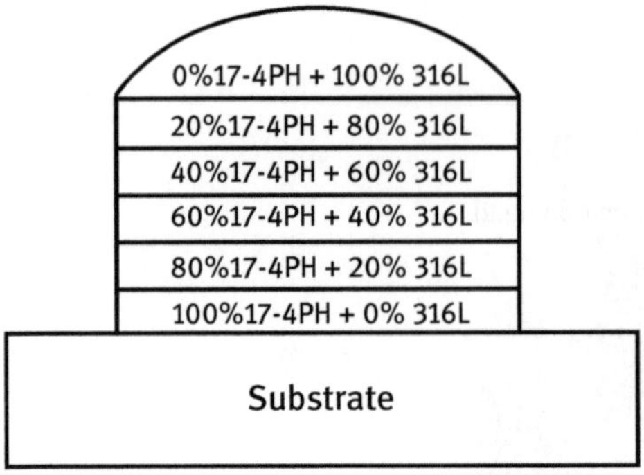

Figure 4.5: Schematic diagram of the FGM.

4.2.4 Specimen characterization

The laser deposited material was sectioned perpendicular to the scanning direction and then metallographically prepared in accordance to the standard preparation for stainless steel [42]. Samples for microscopy studies were chemically etched using Kalling's No.2 etchant (5 g copper(II) chloride, 100 ml hydrochloric acid, 100 ml ethanol). The microstructure was investigated using Tescan scanning electron microscope (SEM). Elemental analysis of etched samples was also investigated using the Oxford Instrument energy dispersion spectrometry fitted in the SEM. The microhardness profile of the sample was measured using Metknon (MH-3) Vickers indenter. The test was carried out as prescribed by the ASTM standard [43]. The measurement was taken from the top surface to bottom of sample at an interval of 0.3 mm. The machine was set to run at load and dwell time of 500 g and 30 s, respectively.

4.3 Results and Discussion

4.3.1 Macrostructure

Figure 4.6 shows the cross section of the fabricated 316L/17-4PH stainless steel composite. As can be seen from the SEM micrograph, the specimen appears to be structurally sound with no evidence of microcracks or pores. Intersecting bowl-shaped features representing melt pool boundaries formed as a result of laser scan and overlap are noticeable. The maximum height of FGM was about 4.8 mm and the average melt pool depth across the substrate was 0.445 mm. The low dilution observed could be attributed to the short interaction time of the laser beam on the substrate due to the high laser scan speed used during processing. Minimal dilution suggests that there is very little mixing between the deposited material and the substrate, hence, reduced contamination with baseline material. Laser-processed

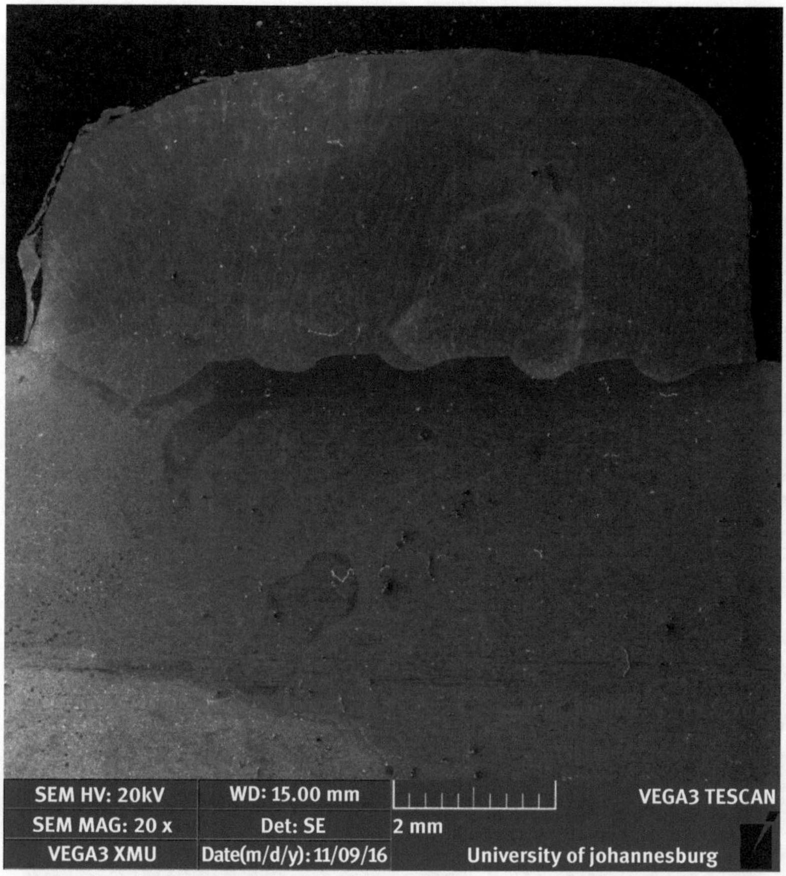

Figure 4.6: SEM cross-sectional image of the FGM.

materials with minimal dilution are desirable compared to high dilution deposition [44, 45]. However, the deposit and substrate interface is the weakest point in the structure, so having a high dilution clad is considered beneficial in certain applications as it ensures a stronger metallurgical bond between the substrate and deposited layer.

4.3.2 Microstructure

Figure 4.7 shows the SEM images taken at different positions on the FGM. Variation in microstructure was observed across the deposition as the compositions from one end of the component to the other. The microstructure changes gradually from a wholly martensitic microstructure at the bottom of the deposit to an austenitic microstructure at the top of the clad. The predominantly martensitic microstructure at the

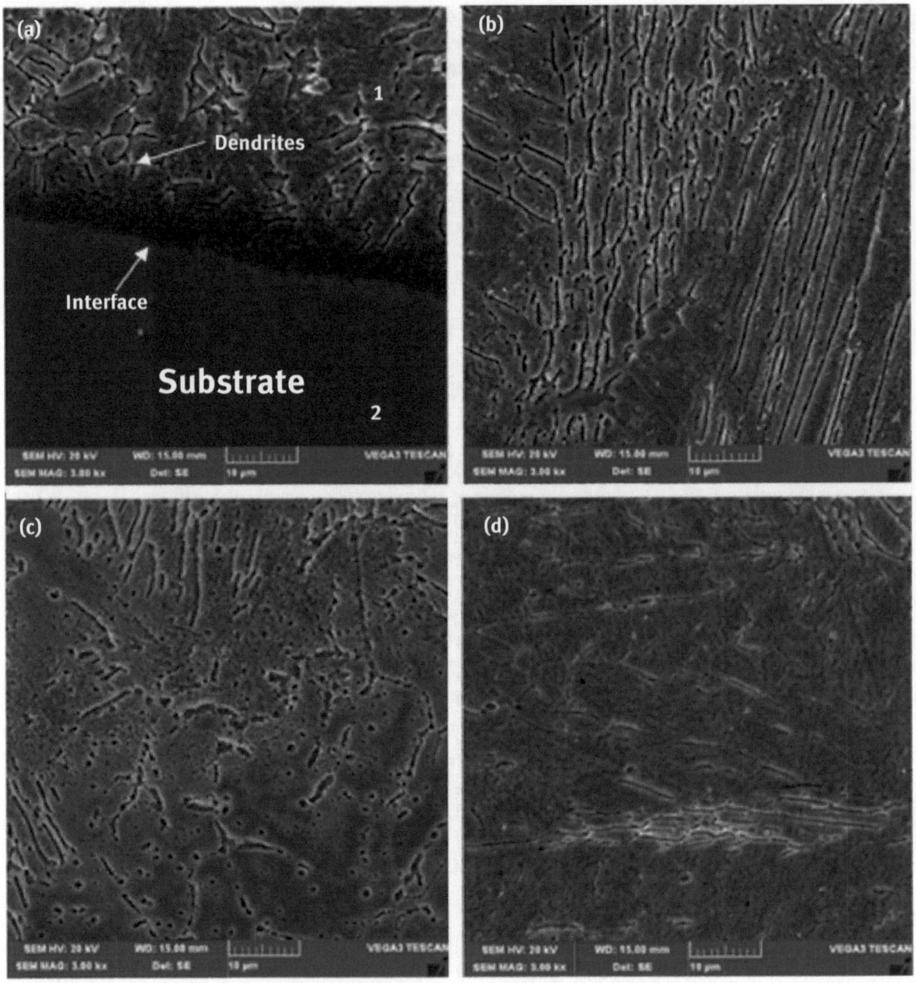

Figure 4.7: Microstructure at different regions of the FGM (a) lowest level, (b) lower level, (c) middle and (d) top.

bottom was expected since dilution was quite low resulting in very minimal mixing between austenitic substrate and martensitic 17-4PH powder. No signs of grain crystallization and heat-affected zone were observed in the vicinity of substrate-deposit interface as shown in Figure 4.7(a). This may be because austenitic stainless steel is generally non-transferable [46]. The region above the interface is characterized dendritic grain structure which grew epitaxially from the substrate. The columnar dendrites observed above the interface are as a result of the high solute concentration of deposited 17-4PH powder which favours dendritic growth. The columnar dendrites are vertically oriented which suggests directionally solidification. Figure 4.8 shows the EDX analysis of spots 1 and 2 depicted in Figure 4.8(a). The

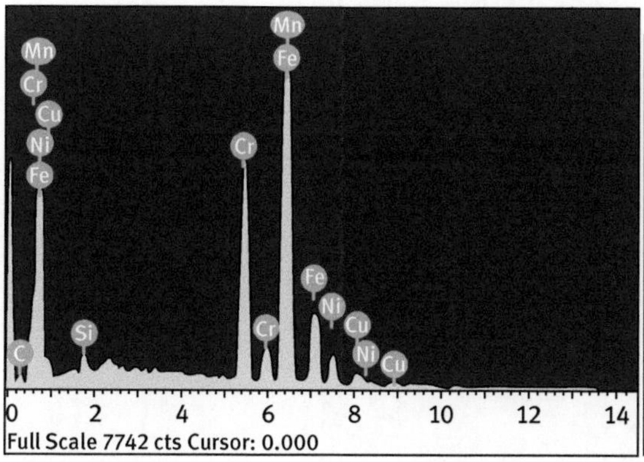

Element	Weight%	Atomic%
C K	3.04	12.58
Si K	0.71	1.25
Cr K	16.41	15.68
Mn K	1.08	0.98
Fe K	70.10	62.37
Ni K	5.82	4.92
Cu K	2.85	2.23
Totals	100.00	

(a)

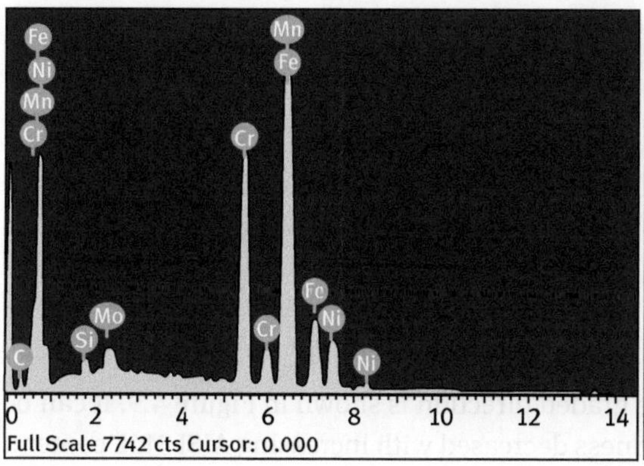

Element	Weight%	Atomic%
C K	3.05	12.68
Si K	0.58	1.03
Cr K	17.49	16.82
Mn K	1.11	1.01
Fe K	66.23	59.31
Ni K	9.46	8.06
Mo L	2.09	1.09
Totals	100.00	

(b)

Figure 4.8: EDX analysis of (a) spot 1 and (b) spot 2.

EDX analysis was carried out to evaluate interdiffusion of elements between the substrate and deposited layer. As shown in Figure 4.8, no transfer of elements appears to have occurred between the substrate and cladded layer, since elements such as Cu and Mo found in 17-4PH powder and 316 substrates, respectively, did not diffuse across both regions.

The rest of the structure also appears to be predominantly dendritic except for the top layer. The lower and middle levels of the FGM shown in Figure 4.7(b) and 4.7(c), respectively, are characterized by columnar grains with and without secondary dendrite arm. The grains also appear to have grown epitaxially from previous layers as opposed to re-nucleating across layers. Thermal gradient increases with height of the deposition, and also the effect of heat accumulation becomes more substantial as more layers are added. These factors favour dendritic growth and grain refinement. The top of the FGM is characterized by fine nodular structure as shown in Figure 4.7(d).

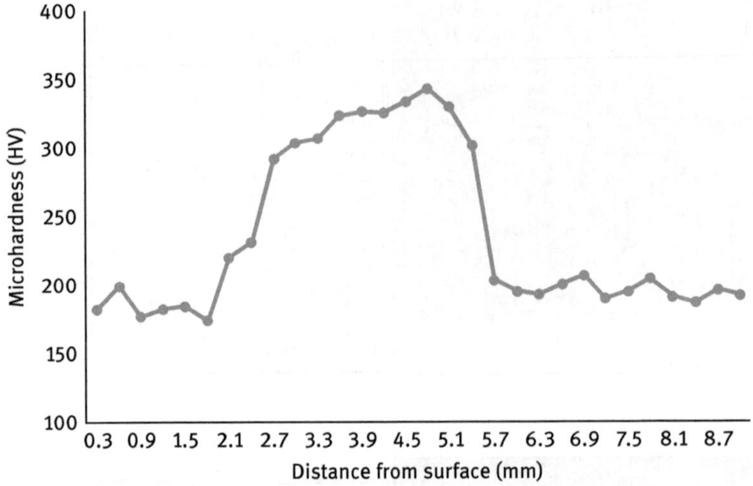

Figure 4.9: Microhardness profile along the depth of the FGM.

The finer grains and nodular structure observed can be accredited to rapid cooling at the surface by the covering gas.

4.3.3 Microhardness

The microhardness profile along the graded direction is shown in Figure 4.9. It can be inferred from the graph that the hardness decreased with increasing AISI 316L content. This was expected since austenite is a weaker phase compared to martensite. The region just above the interface comprising about 100% 17-4PH had the highest hardness value of HV 347. As can be seen from the microhardness profile, there is very little variation in the hardness values between the top layer which is 100% AISI 316L (183 HV) and wrought 316 substrates (186 HV). The similarity in the microhardness value of the top layer is indicative of the non-hardenability of austenitic stainless steel.

4.4 Conclusions

FGMs are a class of advanced materials with exceptional properties and characteristics. The FGM concept allows for tailoring of properties for diverse applications. Compositionally graded AISI 316L and 17-4 PH stainless steel composite was successfully manufactured using LMD technique. The FGM was defect free and showed no evidence of porosity. The microstructural analysis conducted revealed that the microstructure varied with depth. A similar trend was observed with microhardness of the structure. The hardness increases with depth, with the highest hardness value recorded at the bottom of the FGM with 100% 17-4PH content.

References

[1] Bunsell AR, Renard J. Fundamentals of fibre reinforced composite materials. Boca Raton: CRC Press, 2005.

[2] Miyamoto Y et al. Functionally graded materials: design, processing and applications. Amsterdam: Elsevier, 1999.

[3] Ilschner B. Structural and compositional gradients: basic idea, preparation, applications. J Phys IV Coll 1993:C7-763–C7-772.

[4] Koizumi M, Niino M. Overview of FGM research in Japan. MRS Bull 1995;20(01):19–21

[5] Jung Y et al. Residual stress and thermal properties of zirconia/metal (nickel, stainless steel 304) functionally graded materials fabricated by hot pressing. J Mater Sci 1997;32(14):3841–50.

[6] Kim J, Kim M, Park C. Evaluation of functionally graded thermal barrier coatings fabricated by detonation gun spray technique. Surf Coat Technol 2003;168(2):275–80.

[7] Birman V, Keil T, Hosder S. Functionally graded materials in engineering. In: Structural Interfaces and Attachments in Biology. Anonymous, Springer, 2013:19–41.

[8] Besisa DH, Ewais EM. Advances in functionally graded ceramics–processing, sintering properties and applications. In: Advances in functionally graded materials and structures. Anonymous, In Tech, 2016.

[9] Bharti I, Gupta N, Gupta K. Novel applications of functionally graded nano, optoelectronic and thermoelectric materials. Int J Mater Mech Manuf 2013;1(3):221–4.

[10] Mahamood RM, Akinlabi ET. Functionally graded materials. 1st ed. Cham: Springer International Publishing, 2017.

[11] Cahn RW, Haasen P, Kramer EJ. Materials science and technology: a comprehensive treatment, Germany: VCH, 1996;8.

[12] Kieback B, Neubrand A, Riedel H. Processing techniques for functionally graded materials. Mater Sci Eng: A 2003;362(1):81–106.

[13] DeHoff R. Engineering of microstructures. Mater Res 1999;2(3):111–126.

[14] Akinlabi ET, Akinlabi SA. Characterization of functionally graded commercially pure titanium (CPTI) and titanium carbide (TiC) powders, in Proceedings of the World Congress on *Engineering*, London, UK, 2015;2.

[15] Mahamood RM, Akinlabi ET. Laser metal deposition of functionally graded Ti6Al4V/TiC. Mater Des 2015;84:402–10.

[16] Wu D, Liang X, Li Q, Jiang L. Laser rapid manufacturing of stainless steel 316L/Inconel718 functionally graded materials: microstructure evolution and mechanical properties. International Journal of Optics, 2010;2010:1–5.

[17] Lin X, Yue T, Yang H, Huang W. Laser rapid forming of SS316L/Rene88DT graded material. Mater Sci Eng: A 2005;391(1):325–36.

[18] Bever MB, Duwez PE. Gradients in composite materials, Materials Science and Engineering, Jan 1972;10:1–8.

[19] Jamaludin SN, Mustapha F, Nuruzzaman DM, Basri SN. A review on the fabrication techniques of functionally graded ceramic-metallic materials in advanced composites. Sci Res Essays 2013;8(21):828–40.

[20] Gandra J, Miranda R, Vilaça P, Velhinho A, Teixeira JP. Functionally graded materials produced by friction stir processing. J Mater Process Technol 2011;211(11):1659–68.

[21] Kumbhar N, Mulay A. Post processing methods used to improve surface finish of products which are manufactured by additive manufacturing technologies: a review, J Inst Eng (India): Ser C 2016:1–7.

[22] ASTM Standard F2792. Standard terminology for additive manufacturing technologies. ASTM Int West Conshohocken. DOI: 10.1520/F2792-12, 2012

[23] Gibson I, Rosen D, Stucker B. Additive manufacturing technologies: Rapid prototyping to direct digital manufacturing, New York: Springer, 2010.
[24] Frazier WE. Metal additive manufacturing: a review. J Mater Eng Perform 2014;23(6):1917–28.
[25] Syed WUH, Pinkerton AJ, Li L. Combining wire and coaxial powder feeding in laser direct metal deposition for rapid prototyping. Appl Surf Sci 2006;252(13):4803–8.
[26] Toyserkani E, Khajepour A, Corbin SF. Laser cladding. Boca Rofon, Florida: CRC Press, 2004.
[27] Mahamood RM, Akinlabi ET, Shukla M, Pityana S. Laser metal deposition of Ti6Al4V: a study on the effect of laser power on microstructure and microhardness. 2013.
[28] Shukla M, Mahamood RM, Akinlabi ET, Pityana S. Effect of laser power and powder flow rate on properties of laser metal deposited Ti6Al4V. Eng Technol 2012;71:1268–72.
[29] Bayode A, Akinlabi ET, Pityana S. Characterization of laser metal deposited 316L stainless steel. Proceedings of the World Congress on Engineering, London, 2016:925.
[30] Mahamood RM, Akinlabi ET. Effect of laser power on surface finish during laser metal deposition process. Proceedings of the World Congress on Engineering and Computer Science, London, UK, 2014.
[31] Akinlabi ET, Mahamood RM, Shukla M, Pityana S. Effect of scanning speed on material efficiency of laser metal deposited Ti6Al4V, 2012.
[32] Mahamood RM, Akinlabi ET, Shukla M, Pityana S. The role of transverse speed on deposition height and material efficiency in laser deposited titanium alloy, 2013.
[33] Pityana S, Akinlabi ET, Mahamood RM, Shukla M. Gas flow rate and powder flow rate effect on properties of laser metal deposited Ti6Al4V, 2013.
[34] Anandkumar R, Almeida A, Vilar R, Ocelík V, De Hosson JM. Influence of powder particle injection velocity on the microstructure of Al–12Si/SiCp coatings produced by laser cladding. Surf Coat Technol 2009;204(3):285–90.
[35] Erinosho MF, Akinlabi ET, Pityana S. Effect of powder flow rate and gas flow rate on the evolving properties of deposited Ti6al4v/Cu composites. In Advanced Materials Research, 2014;1016:177–182.
[36] Senthilkumaran K, Pandey PM, Rao P. Influence of building strategies on the accuracy of parts in selective laser sintering. Mater Des 2009;30(8):2946–54.
[37] Mahamood RM, Akinlabi ET, Shukla M, Pityana S. Characterizing the effect of laser power density on microstructure, microhardness, and surface finish of laser deposited titanium alloy. J Manuf Sci Eng 2013;135(6):064502.
[38] Sahasrabudhe H, Harrison R, Carpenter C, Bandyopadhyay A. Stainless steel to titanium bimetallic structure using LENS™. Addit Manuf 2015;5:1–8.
[39] Lin X, Yue T, Yang H, Huang W, Microstructure and phase evolution in laser rapid forming of a functionally graded Ti–Rene88DT alloy. Acta Mater 2006;54(7):1901–15.
[40] Noecker F, DuPont JN. Functionally graded copper-steel using laser engineered net shaping process. Proceedings of the Solid Freeform Fabrication Symposium, 2002:231–8.
[41] Carroll BE, Otis RA, Borgonia JP, Suh J, Dillon RP, Shapiro AA, Hofmann DC, Liu Z, Beese AM. Functionally graded material of 304L stainless steel and Inconel 625 fabricated by directed energy deposition: characterization and thermodynamic modeling. Acta Mater 2016;108:46–54.
[42] Weidmann E, Guesnier A, Taylor B. Application notes: metallographic preparation of Stainless Steel. Available at: www.struers.com, 2015.
[43] ASTM Standard E384-11E1. Standard test method for Knoop and Vickers hardness of materials. ASTM International, West Conshohocken. DOI: 10.1520, 2007.
[44] Steen W, Mazumder J. Laser material processing. Laser Material Processing, 4th ed. London: Springer Science & Business Media, 2010.
[45] Abboud J, Rawlings R, West D. Functionally gradient layers of Ti–Al based alloys produced by laser alloying and cladding. Mater Sci Technol 1994;10(5):414–20.
[46] Esfahani MN, Coupland J, Marimuthu S. Microstructure and mechanical properties of a laser welded low carbon–stainless steel joint. J Mater Process Technol 2014;214(12):2941–8.

M. H. Alaaeddin, S. M. Sapuan, M. Z. M. Yusoff, E. S. Zainudin, and Faris M. AL-Oqla

5 Natural fiber composites as functionally graded materials for advanced applications

Abstract: The characterization of natural fiber composites (NFCs) has led to a new era for the utilization of its properties in multiple tasks and a variety of applications. This work aimed at investigating the prospects of NFCs to be as functionally graded materials (FGMs) for the advanced applications considering methodologies for material selection. However, the recent development of NFCs resulted in the emergence of more creative methods to utilize NFs within a countless number of applications; the necessity of these applications derives from the persistent need to be used in all aspects of life, hence, categorizing its characteristics and properties ensuring wider and more systematic classification considering a gradual variation of natural fiber (NF) structure over certain measurements and reinforcements. As a result, NF-reinforced polymer or co-polymer showed respectable results for a considerable number of applications, which can be workable for advanced applications, for example, automotive industry, constructions, and renewable energy production. The review introduced a new concept consisting of FGM selection possibilities and potentials. In contrast, more in-depth studies and empirical researches are required to improve the macrostructure and microstructure of NFC-reinforced polymer to enhance its toughness, strength, and flatness and to improve its material functionality where required.

Keywords: natural fiber composites (NFCs), functionally graded materials (FGMs), materials selection (Ms), advanced applications

5.1 Introduction

In recent times, natural fiber composites (NFCs) have demonstrated valuable exploitations within important and disparately needed applications considering a variety of different species of natural fiber (NF), for example, kenaf, jute, hemp, flax, and sisal [1, 2]. However, functionally graded materials (FGMs) have great potential as a substitute material that operates in several conditions [3, 4]. Therefore, graded materials are embedded within countless objects and entities; skin and bones have functional grading, chemically and spatially compatible, and advanced structural materials. In addition, a number of FGMs are naturally occurring and some engineered by humans [5, 6]. Taking an example at the level of both NFCs and FGMs, bamboo is an NF that is widely used in certain constructions and as a composite; it reflects the concept of FGMs with a complicated microstructural shapes material distribution [7].

https://doi.org/10.1515/9783110545104-005

5.2 NFCs for potential use as FGM

The utilization of NF became exceedingly important for a number of applications. Recently, NFCs became highly valuable, environmentally attractive, and controllable materials that compete with a number of other materials used in the latest fittings [1, 8–13]. Several researches investigated the properties and applications of NFCs to a certain extent since the early 1900s. In the late 1980s, more attention has been given due to its compatible properties to certain applications, reinforcing or strengthening production and low cost [14]. More importantly, NFCs asserted to have a high biodegradability and recyclability, which reflects its importance to environment besides the enhanced energy recovery, acceptable specific strength with low weight, low density, and high toughness [15, 16]. In technology, NFCs have proved to be an interesting choice for a number of technological applications; it is a broadly applied fiber in composite technology and increasingly considered for polymer–matrix composites as reinforcements [17]. In addition, the chemical, mechanical, and physical properties and the inherent interaction of fibers and matrices usually determine the final attributes and performance of NFCs. Table 5.1 shows the property range on analytical average values of six different natural fiber categories.

Biological organisms degrade NFs as the carbohydrate polymers are recognized in the cell wall, the lignocelluloses component bared which processed in a photochemical degradation that instigated by ultraviolet light. However, the anticipated resistance to biodegradation and ultraviolet light is possible to be polished through bonding chemicals to the cell wall or adding polymers to the cell matrix itself [2, 10, 12, 14, 19, 20]. In addition, NFs are environmentally superior and interact well with most performance metrics; they integrate as material composition with different components as long as they are properly selected as multifunctional materials for the intended applications [9, 14, 15, 20, 21].

Table 5.1: Property range and average values of selected fibers [18].

Fiber type	Density (g/cm^3)	Tensile strength (MPa)	Tensile modulus (GPa)	Elongation to break (%)
Jute	1.4	560	43.0	1.4
Coir	1.31	162.5	4.4	33.2
Kenaf	1.4	576.5	33.75	2.1
Date palm	1.05	186.0	7.25	10.05
Hemp	1.45	585.0	56.75	2.25
Oil palm	1.13	164.0	1.85	21.0

5.3 Material selection in NFCs as FGMs

In modern technology, three main pillars are imperatively considered to produce practical and useful applications, which are materials, energy, and modern science, the increased interdisciplinary interactions that subjected materials science to a rapid development [21, 22]. Natural biomaterials usually possess the general concept of FGM structuring, which qualifies the two concepts to integrate with each other and reform a new concept that simultaneously satisfies the many requirements in certain applications and improves the methodologies of material selection [22, 23]. A beneficial evaluative criterion is required to select the best materials for specific applications or manufacturing processes. Numerous issues and related criteria have to be taken into account when selecting materials to improve applications or optimize operations [19, 20]. The concept of functionally graded composite materials (FGCMs) reflects a change in microstructure and composition that occurs incessantly with position. However, the main challenge is to make a component that contains irreconcilable properties such as structural toughness at low temperatures and light hardness at high temperatures [21]. Material selection for FGMs has become a complex process, which requires properties to be demonstrated by design and structure that necessitates advanced material selection paradigm considering a simultaneous design of material composition and structure; for NF or synthetic materials a complete parameterization and understanding for the generative process is required to fulfill the desired class of material structure [24]. A considerable number of factors need to be taken into account in order to select the most fitting material-related features in the selection process to maximize quality of operating, design and durability, and to minimize cost and reduce damage [20, 23–25]. These factors have physical, mechanical, magnetic, electrical, and industrial properties [20, 23–26]. Meanwhile, like any other development process of a product, there are two stages of material selection, which are alternative generation and alternative selection, and both are important to fulfill the process. Subsequently, the material selection flow illustration provides a complete selection process to identify material component and performance specification as shown in Figure 5.1.

Designing and remanufacturing products involve integrated manufacturing considerations into various aspects to bring these products in its best and final form. The intrinsic values of the meant components can be well maintained and kept up as long as the quality of the product or application assured in the manufacturing process, and considering the factors that contribute to this process, for example, material selection process, joint selection, and structure configuration, this can be achieved considering a comprehensive list of criteria for material evaluation relevant to the manufacturing product using a multicriteria decision-making including different software methodologies [22, 24, 26, 27–29].

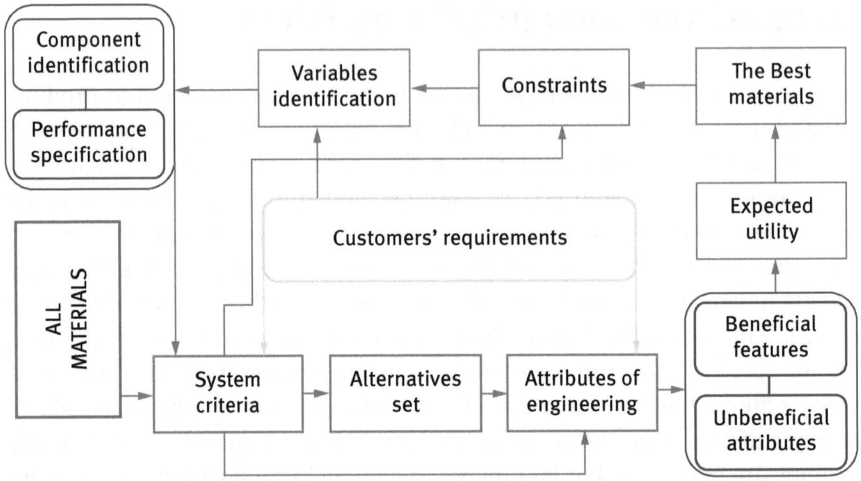

Figure 5.1: Material selection decision-based flow diagram [27].

5.4 Methods of materials selection that can be used to select NFCs as FGMs

Finding new and more realistic methodologies for material selection is highly encouraged in engineering. To avoid uncertainties and time consuming, there are different processes, charts, and techniques to select the most appropriate materials with various methods and capabilities such as the analytical hierarchy process (AHP) included in Expert Choice software and expert systems as analytical methods, TOPSIS method, Pugh selection method, Ashby charts, Java-based material selection, Cambridge material selector (CMS), knowledge-based system (KBS), computer-aided material selector (Plascams) (CAMPUS), digital logic technique, artificial neural network (ANN), and quality function deployment (QFD) [24, 28–31]. Table 5.2 shows the different methods of material selection comparing the conventional methods to the new computerized methods.

The material selection in FGMs in advanced applications demands the development of advanced composite selection methods that can withstand high gradients of different requirements, for example, thermal resistivity, expansion, fabricability, conductivity, resistance to erosion, durability, water proofing, macro- and micro-structuring, and many other related issues [27, 31–34]. Figure 5.2 shows sample property charts of Ashby visualizing and communicate material properties for material selection process. The sample is taken from CES selector (GRANTA Material Intelligence) as cited.

Table 5.2: The different methods of material selection comparing the conventional methods to the new computerized methods [23, 31–35].

Conventional methods	New computerized methods
Cost per unit property approaches	Artificial intelligence methods
Materials in products methods of selection	Genetic algorithm and neural network
Exploratory work, observation, sampling, testing, and analysis	programming and software analysis (TOPSIS, AHP, CMS, KBS Ashby)
Survey, enumerating, and case study	Simple additive weighting method (SAW)
Screening and experience evaluation method	Multiple attribute decision-making (TOPSIS, fuzzy, MADM) and MODMM
Experiments and observational selection methods	Optimization methods software (CAD/CAM), genetic algorithm, MP

5.5 Computerized material databases

In computer-aided systems, computerized material databases have an imperative role for material selection process. It simply supports specified numeric values, for example, weldability, formability, machinability, and availability. It also provides more accurate illustrations over the class of materials, forms, designation, and composition [36]. It is advisable to use materials simulation software methods for FGMs to improve the process of selection [36–40]. There are several important approaches and useful methods to obtain the compositional gradient of FGMs, including gas based, liquid phase, solid phase, and methods that can be utilized to obtain chemically or physically tailored properties [37, 38]. However, ideal material selection is an essential function to design all products; the methods have become interestingly advanced and consider digital logic and sophisticated software applications [38]. In addition, making calculations for the values assumed by different functions to quantify impacts and handling issues in the components of materials represent interesting values to the advanced selection process of materials including computer-controlling systems, to handle the various issues and structures [37, 39]. Therefore, having a multicriteria selection model extracted from the fundamental parameter concept of material selection methods considering the advanced and computer-based systems contributes to new ideal forms of optimization and enriches any progressive change in materials requirements [40]. Combining two or more different methods in material selection processes improves reliability and provides ideal solutions in certain cases, for example combining TOPSIS and AHP as method-based approach for nontraditional machining processes selection [41–43]. Therefore, in order to stand for the required specifications in the process of selection, it is important to demonstrate integrity in data collection and arranging database with no missing information to represent the case properly.

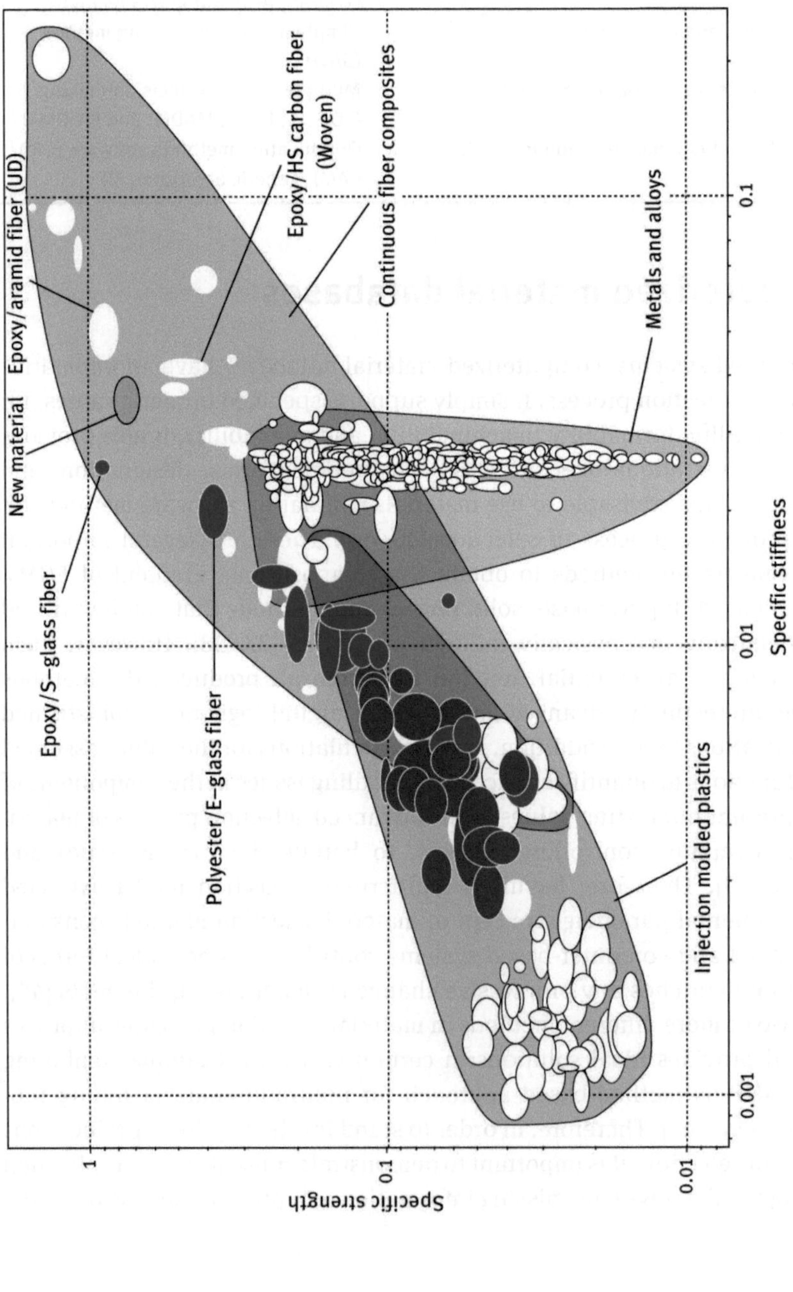

Figure 5.2: Comparing a new composite material with the performance of other composites, plastics, and metals [35].

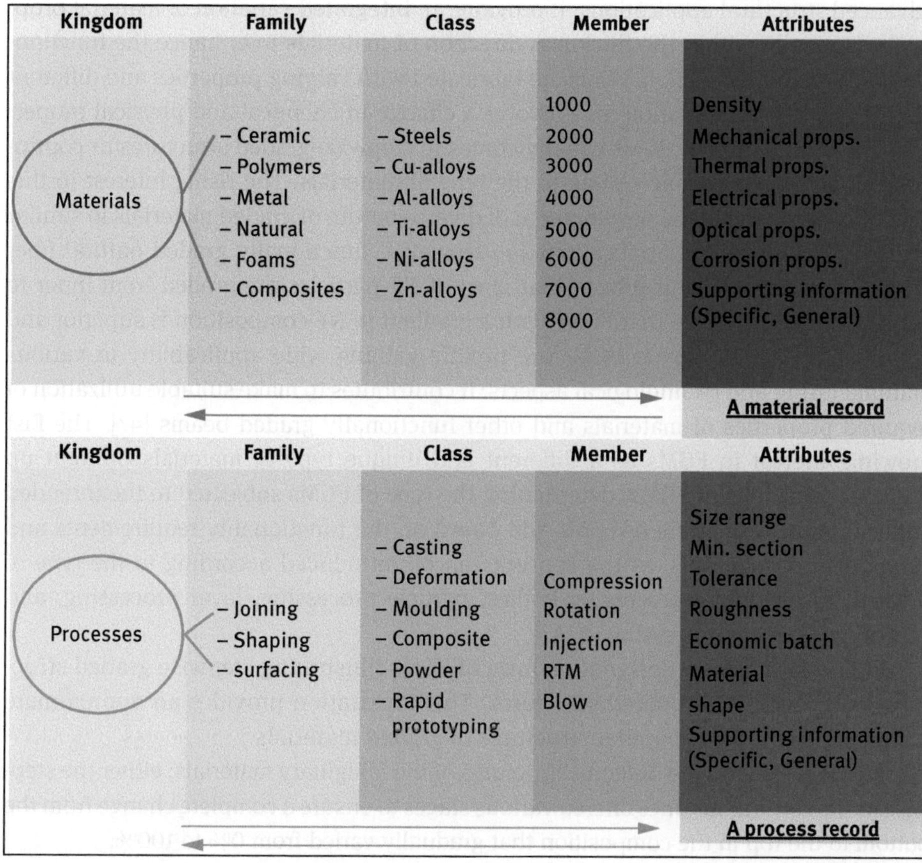

Figure 5.3: An example for the organizational process of structured data for material selection [42].

The hierarchical structure in Figure 5.3 illustrates a good strategy for both "material" and "process" organizing the structured data; the tree form shows a comprehend collection for data considering materials and processes to be joined with a complete analytical process [42].

5.6 Functionally graded materials

FGMs are different and unique from other types of material; it is widened or developed materials that categorized by variation in its characteristics according to certain required function [43]. However, the definition of FGMs emerged due to the need to innovated methods of fabrication for certain type of material composites that made up of various parts and elements. Simply, it is a new generation of engineered materials [44]. The utilization process of FGMs developed to be necessarily advisable for advanced structural applications; it provides an integrated variation of material properties especially within the thickness direction of materials to enhance the functionality of these materials [45]. FGMs are fabricated with varying properties and different characteristics; the variation may involve a change in chemical and physical properties. The designation of FGMs usually comes in stepwise-graded structures or continuous-graded structures according to the type of materials. The rising interest in this type of material led to the development of diverse nature of graded materials in similar or dissimilar areas and applications [4]. However, functionally graded natural fiber (FGNF) as a new concept proved that most of NF particles are graded from inner to outer surface; it appears that FGNF with a gradient in NF composition is superior and remarkable [46]. Moreover, FGMs are rapidly gaining wide applicability in various manufacturing and technological aspects. It contributes to make suitable utilization of required properties of materials and other functionally graded beams [47]. The fast growing interest in FGMs as a different and unique type of materials brought up different categories of FGMs, determining the type of FGMs subjected to the intended application in wide scale of FGMs and based on the functionality requirements and component of materials, as the concept can be introduced according to the type of process, for example, processing bodies, particle processing, layer processing, and centrifugal casting processing [4, 47].

Figure 5.4 shows a designation form of FGMs illustrating stepwise-graded structures and continuous-graded structures. This illustration provides an approximate imaginary to the most popular structures of graded materials.

Figure 5.4 shows the different structures of the imaginary materials; either the stepwise or the continuous ones offered various stages to ensure a complete change from the bottom to the top in the composition that gradually varied from 0% to 100%.

Jackson [48] indicated some constraints of the composition on the volume fraction resolution in FGM model considering the vectors of volume fractions of the base materials. The compositions introduced in a gray-valued voxel model are illustrated in Figure 5.5.

However, to guarantee a certain accuracy in composition representation with a minimum resolution for the volume fraction of tested material, the same study considered the equation of $n\lambda = \left\lceil \frac{1}{2\epsilon_m} + 1 \right\rceil$.

As for the constraint based on discontinuities in composition, the discontinuity in composition arises considering $\boldsymbol{\pi}$ and $(\boldsymbol{m} \times (\mathbf{x}))$ occupied by the voxel, the differing composition if $x - xi + y - yj + z - zk < \beta a$, then

$$m \times (x, y, z) = \left\{ \frac{\begin{bmatrix} 1.0 \\ \square \\ 0.0 \end{bmatrix}}{\begin{bmatrix} 0.0 \\ \square \\ 1.0 \end{bmatrix}} \right. \qquad [5.48]$$

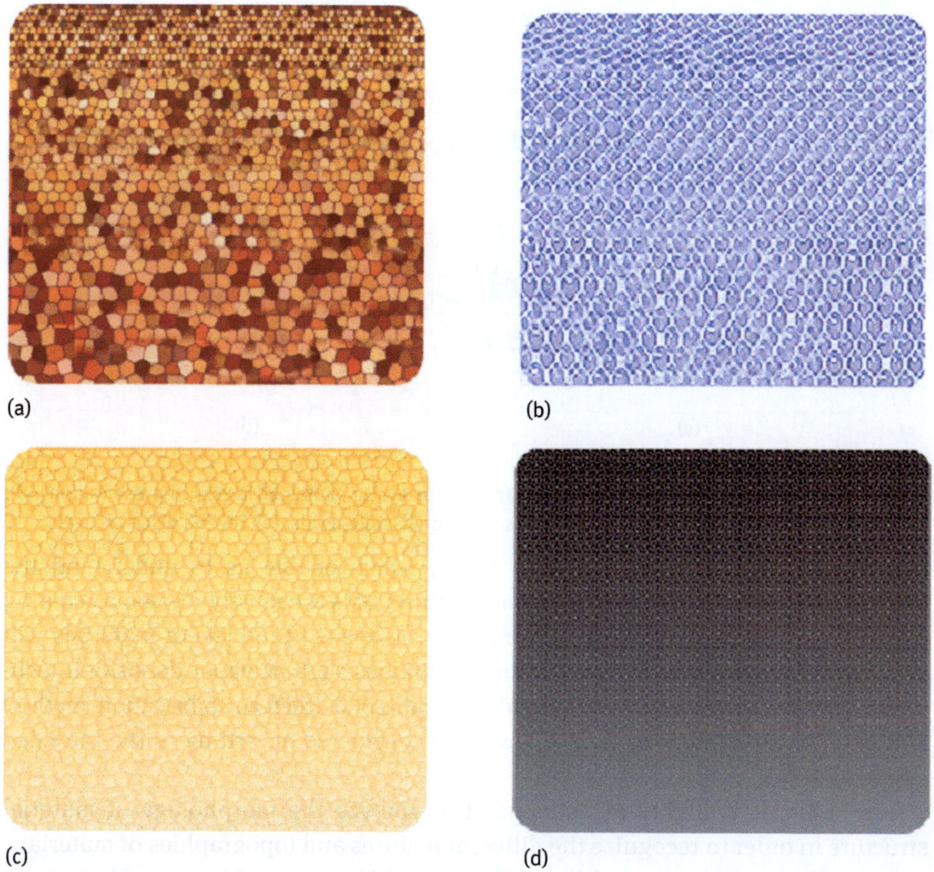

Figure 5.4: Designation of FGMs shows an illustration to stepwise-graded structures: (a) a stepwise-graded structure (segmented), (b) a stepwise-graded structure (radial), (c) a continuous-graded structure (classic gradient), and (d) a continuous-graded structure (filled gradient).

5.7 FGM modeling

In order to identify the capability to model graded composition, it is necessary to consider that this capability requires a demonstration including the subregions of uniform materials along with the decomposing models. However, there are several approaches for the modeling of FGMs, for example, finite element modeling (FEM), voxel-based modeling, and modeling methods that generalized in different operations [48]. However, it is important to realize that FGMs are indeed different from the conventional composites or one-function materials. The mechanical properties of FGMs, for example, shear modulus of elasticity, Young's modulus of elasticity, material density, and Poisson's ratio vary in preferred directions [5]. There are a

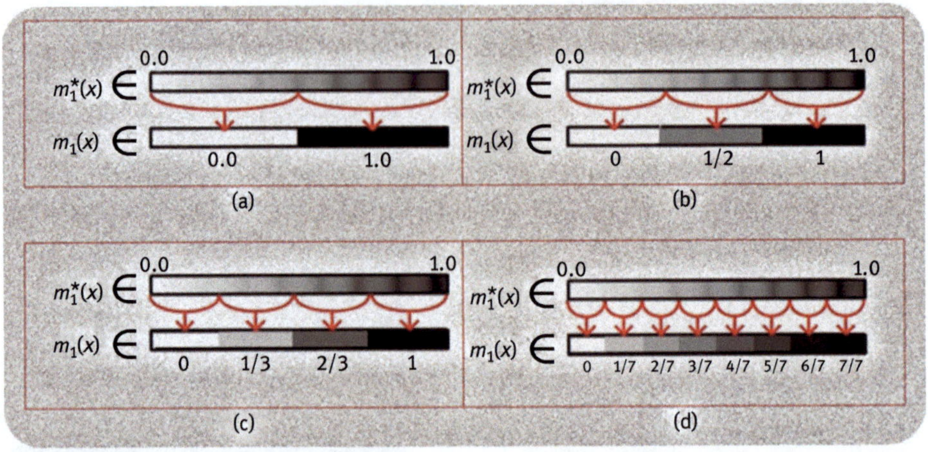

Figure 5.5: Thresholding of continuous grading to discrete levels maintained in voxel representation [48]. (a) Two discrete levels. (b) Three discrete levels. (C) Four discrete levels. (a) Several discrete levels.

number of processing techniques of FGMs, which enable us to understand the methods of functionality to such materials; thin FGMs are mostly formed as coating on surfaces, having a dimensional range for the needed products or materials will ease the selection method of processing techniques such as vapor deposition technique, powder metallurgy, centrifugal method, solid freeform fabrication method and other more interesting techniques to be considered according to the intended application [43].

In FGMs, it is highly recommended to analyze the morphology of material structure in order to recognize the different features and topographies of materials. To describe the structure of lignocellulose of fibers, consider jute fiber as an example anticipating that other fibers such as flax, kenaf, abaca, coir, sisal, and hemp could have similar general tendencies, whereas the differences will be noted in the characteristics and composition. Figure 5.6 shows the detailed morphology of jute fiber.

Jute is notorious as a multicellular fiber, and the cell wall consisting of a primary and secondary cell wall is divided into three layers. The primary cell wall is shaped with oriented microfibrils, whereas the secondary cell wall, that is, the second layer, occupies the largest volume compared to the other two layers with three hierarchical microstructures, which are microfibril, macrofibril, and micelle.

Generally, the composition of the cell walls determines the evolutionary diversity and multiplicity. However, the stem of the jute plant is a hollow woody core; a layer of blast contains the long jute fiber covering the stem. The process to separate the jute fiber from the woody stem is called retting. The number of cells (the ultimate cell could be from 2.5 to 3.0 mm long) would vary along the length of jute fiber element [49].

Consequently, bamboo modeling process with graded finite elements has important considerations to be taken into account; the variation range of materials within the

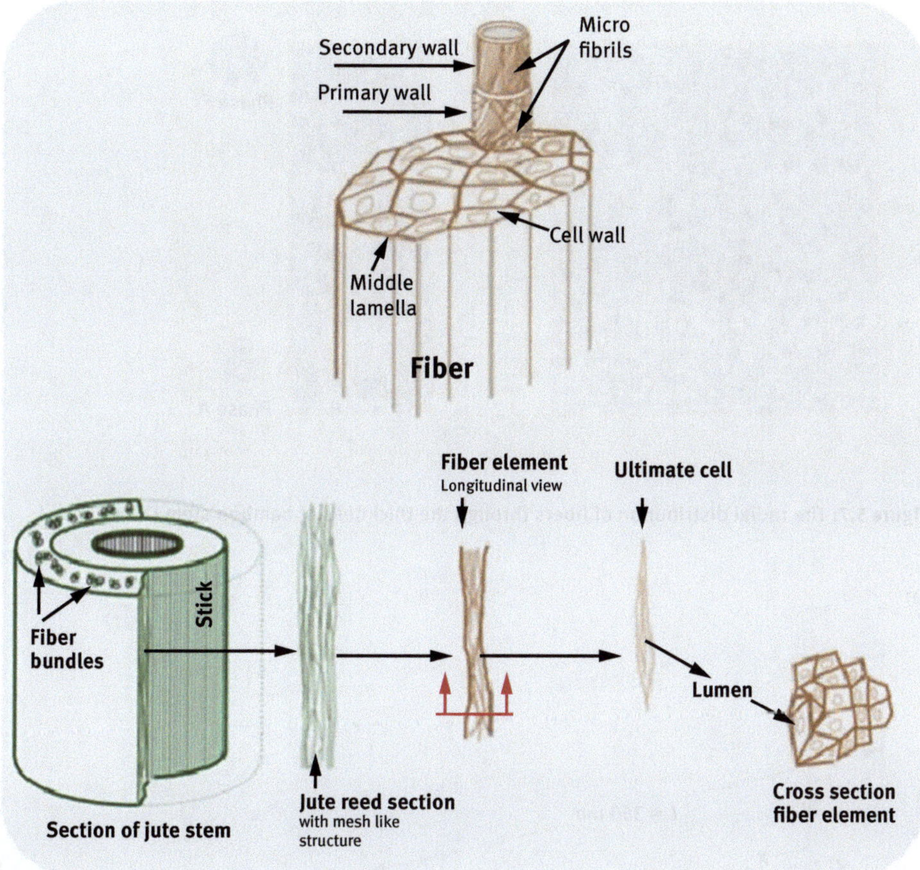

Figure 5.6: Jute stem and fiber morphology and structure [49].

process is an important consideration. Moreover, the conventional finite element formulation properties are presumed to be persistent within each element. Therefore, an incessant distribution of materials could be approached by piecewise-constant element in the modeling process. Figure 5.7 shows a cross section of radial distribution of fibers within the thickness of bamboo stem [7].

Figure 5.8(a) shows the FEM model constructed for stress loading, which includes 7,380 20-noded brick finite elements and 33,794 nodes, whereas Figure 5.8 (b) shows a view of mesh discretization at the end of cell.

Figure 5.8(b) shows half dimensions assumed for finite element interconnects. To equate the axisymmetric stiffness tensor, ***r*** signifies the position in the thickness of the cell wall beginning at the internal surface, whereas ***t*** is the thickness of the anticipated cell wall: [7].

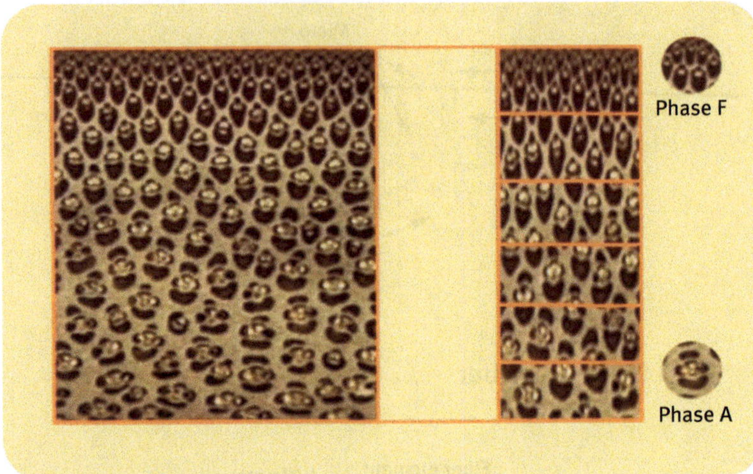

Figure 5.7: The radial distribution of fibers through the thickness of bamboo stem [7, 50, 51].

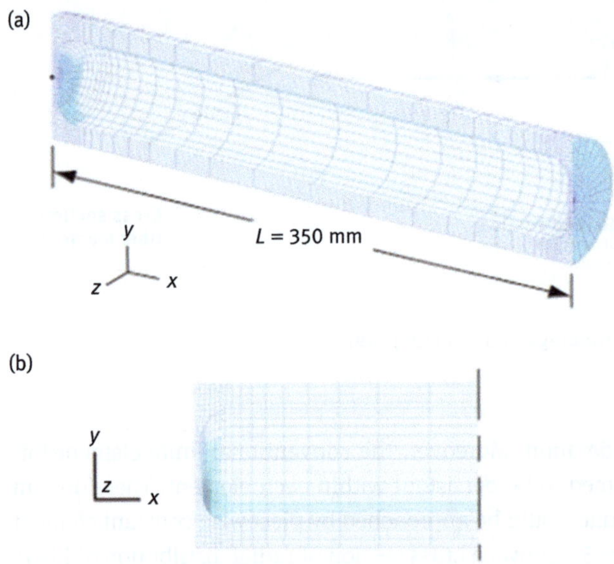

Figure 5.8: Bamboo cell mesh discretization: (a) FEM model in bamboo and (b) a view of mesh discretization at the end of cell [7].

$$E(r)=\frac{(1-V)E\ (r)}{(1+v)(1-2v)}\begin{bmatrix} 1 & \frac{v}{1-v} & \frac{v}{1-v} & 0 \\ \frac{v}{1-v} & 1 & \frac{v}{1-v} & 0 \\ \frac{v}{1-v} & \frac{v}{1-v} & 1 & 0 \\ 0 & 0 & 0 & \frac{1-2v}{2(1-v)} \end{bmatrix}$$

Consequently, the microstructure nature of bamboo varies from volume to another in the same component; the region with high density is called sclerenchyma. However, the sclerenchyma is composed of cellulose microfiber. These groups of fibers are accountable for the bamboo strength, whereas the veins manage the transport of sap from the soil to all parts as the cellulose microfibers that surrounded the veins keep them also straight. However, the lignin that surrounded the vascular bundles named as parenchyma; it does act as a matrix and affords the weak part of the bamboo composite.

However, in order to evaluate the mechanical behavior of composite materials using the rule of mixtures considering a group of equations; these equations provide values for the mechanical properties of composites depending on the volume fraction of the constituents, mechanical properties, matrix, and fibers. The following equation example shows the Young's modulus for a composite:

$$E_C = E_f V_f + E_m V_m = E_f V_f + E_m \left(1 - V_f\right)$$

Equations consider modifying in order to reflect the volume fraction variation with thickness as FGM. However, the equations that contemplate the volume fraction variation along the **x**-axis in the transversal radial direction of the bamboo, which can be assessed through the image analysis methods:

$$E_C = f(X) = E_f V_f(X) + E_m \left(1 - V_f(X)\right) \qquad [5.51]$$

However, consider cotton as an addition example; cotton is one of the large sources of cellulosic fibers besides other types of fibers such as jute, flax, sisal, hemp, and curaua. The dimensional arrangements of unit cells within a fiber reflect on the approved structure that influences the properties of the fiber to match with meant function, which directly affects the polymeric composite properties.

Overall, in plant fiber structure, the single cell wall of each fiber consists of numerous layers starting from the middle lamella, thin primary and secondary walls, and the secondary wall splits into three different segments (secondary wall (S1), middle secondary wall (S2), and the last wall which is the internal secondary wall (S3)). Moreover, the primary consists of disordered arrangements of cellulose fibrils located in a matrix of pectin, hemicellulose, lignin, and protein. S1 contains crystalline cellulose microfibrils arranged in a spiral form and S2 determines the mechanical properties of the fiber. Refer to Figures 5.6 and 5.9 for further details. The fiber nature and strength could be helpful in determining the fiber properties for certain applications. [52].

Table 5.3: Overview on the applications of natural fiber [53].

Fiber	Application
Jute fiber	Transport, roofing systems, building panels, frames, packaging, geotextiles, and boards for certain applications.
Coir fiber	Building panels, storage tank and packaging, helmets and postboxes, voltage stabilizer cover, paper weights, ropes, and many other applications.
Kenaf fiber	Packaging materials, insulations, cases, and materials that absorb oil and liquids, and other applications.
Sisal fiber	Constructions, roofing systems, manufacturing of paper and pulp.
Flax fiber	Panels, decking, railing systems sports, cases, and constructions.
Hemp fiber	Construction products, furniture, textiles, and geotextiles, paper and packaging, and many other applications.
Cotton fiber	Textile and yarn, furniture industry, goods, and cordage.
Bagasse fiber	Constructions (frames and panels), railing systems, and others.
Palm fiber	Constructions, roofing, panels, frames, sliding, and structural insulated panel building systems.
Wood fiber	Window frame, panels, door shutters, decking, railing systems, wooden blades, and fencing and many other applications.

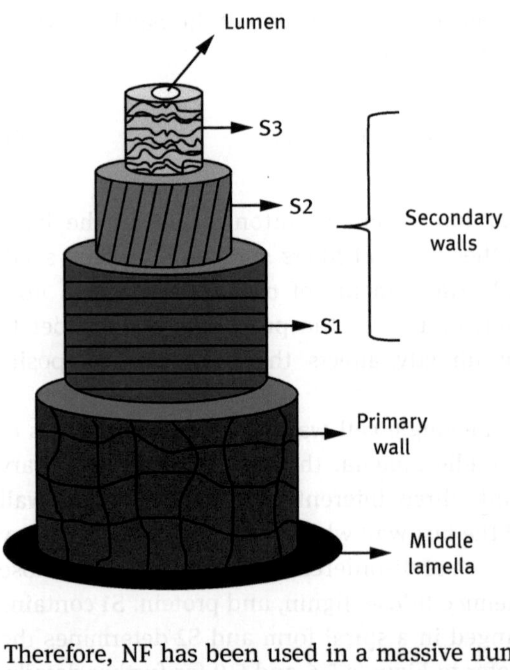

Figure 5.9: A schematic illustration of the plant fiber structure [52].

Therefore, NF has been used in a massive number of applications due to its varying properties. Recently, NF reinforcement or polymer composites is growing rapidly in numerous engineering fields such as automotive industry, constructions, textiles, proofing systems, sports, packaging, furniture, aircraft industry, transportation, and many other applications. Table 5.3 shows the NFC utilization in a wide range of industrial applications.

5.8 Conclusions

There are different scenarios to understand the relationship between NFCs and FGMs based on the targeted function or intended purpose. FGMs are one of the smartest types of materials; they are formed or built for a specific function in a particular part that works in a larger system. Thus, utilizing NF in a composite or a structure that contains on a spatial gradation for a specific function in progressive fields, for example, medical equipment, automotive industry, photovoltaic applications, wind turbines, aerospace, infrastructure, and telecommunication will open further prospects considering the wide range of different shapes and properties that NF has. In addition, this work introduced a new concept that combines FGMs with material selection, which will be reviewed in more detail in future endeavors. On the other hand, the gradual change in the composition or the structure of FGMs in the volume depends on the mechanism of the intended function, which may sometimes be not necessarily binding on the use of NFCs as long as the last do not fit into this functionality.

Acknowledgments: The authors wish to acknowledge the Ministry of Higher Education, Malaysia for the research grant HiCoE project vote number 6369107, and the Universiti Putra Malaysia for facilitating the necessary means to accomplish this work.

References

[1] Al-Oqla FM, Sapuan MS. Natural fiber reinforced polymer composites in industrial applications: Feasibility of date palm fibers for sustainable automotive industry. J Clean Prod 2014;66: 347–54.

[2] Sgriccia N, Hawley CM, Misra M. Characterization of natural fiber surfaces and natural fiber composites. Compos Part A Appl Sci Manuf 2008;39(10):1632–7.

[3] Tarlochan F. Functionally graded material: a new breed of engineered material. J Appl Mech Eng 2013;2(2).

[4] Mahamood RM, Akinlabi ET. Functionally graded materials. Springer, 2017.

[5] Shahistha AC, Binol V, Anjali B. A review on functionally graded materials. Int J Eng Sci 2014; 3(6):90–101.

[6] Udupa G, Shrikantha Rao S, Gangadharan VK. A review of carbon nanotube reinforced aluminium composite and functionally graded composites as a future material for aerospace. Int J Mod Eng Res 2014;4(7):13–22.

[7] Silva EC, Walters CM, Paulino HG. Modeling bamboo as a functionally graded material: lessons for the analysis of affordable materials. J Mater Sci 2006;41(21):6991–7004.

[8] Chapple S, Anandjiwala R. Flammability of natural fiber-reinforced composites and strategies for fire retardancy: a review. J Thermoplast Compos Mater 2010;23(6):871–93.

[9] Ahmad F, Choi SH, Park KM. A review: natural fiber composites selection in view of mechanical, light weight, and economic properties. Macromol Mater Eng 2015;300(1):10–24.

[10] Barkoula MN, Alcock B, Cabrera ON, Peijs T. Fatigue properties of highly oriented polypropylene tapes and all-polypropylene composites. Polym Polym Compos 2008;16(2):101–13.

[11] O'Donnell A, Dweib AM, Wool PR. Natural fiber composites with plant oil-based resin. Compos Sci Technol 2004;64(9):1135–45.

[12] Ku H, Wang H, Pattarachaiyakoop N, Trada M. A review on the tensile properties of natural fiber reinforced polymer composites. Compos Part B Eng 2011;42(4):856–73.

[13] Thakur KV, Thakur KM, Gupta KR. Review: raw natural fiber–based polymer composites. Int J Polym Anal Charact 2014;19(3):256–71.

[14] Nyström B. Natural fiber composites : a review. Eng 2007;15(March):281–5.

[15] Mohanty KA, Wibowo A, Misra M, Drzal TL. Effect of process engineering on the performance of natural fiber reinforced cellulose acetate biocomposites. Compos Part A Appl Sci Manuf 2004;35(3):363–70.

[16] Yusoff M, Salit SM, Ismail N, Wirawan R. Mechanical properties of short random oil palm fibre reinforced epoxy composites. Sains Malaysiana 2010;39(1):87–92.

[17] Norhidayah HM, Hambali AA, Yuhazri MY, Zolkarnain M, Taufik, Saifuddin YH. A review of current development in natural fiber composites in automotive applications. Appl Mech Mater 2014;564(October):3–7.

[18] Al-Oqla FM, Salit SM, Ishak RM, Aziz AN. Combined multi-criteria evaluation stage technique as an agro waste evaluation indicator for polymeric composites: date palm fibers as a case study. BioResources 2014;9(3):4608–21.

[19] Saheb ND, Jog PJ, Nabi Saheb D, Jog PJ. Natural fiber polymer composites : a review. Adv Polym Technol 1999;18(4):351–63.

[20] Joshi VS, Drzal TL, Mohanty KA, Arora S. Are natural fiber composites environmentally superior to glass fiber reinforced composites? Compos Part A Appl Sci Manuf 2004;35(3):371–6.

[21] Udupa G, Rao SS, Gangadharan VK. Functionally graded composite materials: an overview. Procedia Mater. Sci 2014;5:1291–9.

[22] Sadollah A, Bahreininejad A. Optimum functionally gradient materials for dental implant using simulated annealing. Simulated Annealing-Single and Multiple Objective Problems. InTech 2012.

[23] Al-Oqla FM, and Sapuan SM. Materials Selection for Natural Fiber Composites. Woodhead Publishing, 2017.

[24] Liu X, Shapiro V. Sample-based design of functionally graded material structures. ASME 2016International Design Engineering Technical Conferences and Computers and Information in Engineering Conference IDETC/CIE, 2016:1–15.

[25] Kaspar J, Baehre D, Vielhaber M. Material selection based on a product and production engineering integration framework. Procedia CIRP 2016;50:2–7.

[26] Mansor MR, Sapuan SM. Concurrent Conceptual Design and Materials Selection of Natural Fiber Composite Products. Springer, 2017.

[27] Das D, Bhattacharya S, Sarkar B. Decision-based design-driven material selection: a normative-prescriptive approach for simultaneous selection of material and geometric variables in gear design. Mater Des 2016;92:787–93.

[28] Fu X, Schuh AC, Olivetti AE. Materials selection considerations for high entropy alloys. Scr Mater 2017;138:145–50.

[29] Yang SS, Nasr N, Ong KS, Nee AY. Designing automotive products for remanufacturing from material selection perspective. J Clean Prod 2017;153:570–9.

[30] Poulikidou S, Schneider C, Björklund A, Kazemahvazi S, Wennhage P, Zenkert D. A material selection approach to evaluate material substitution for minimizing the life cycle environmental impact of vehicles. Mater Des 2015;83:704–12.

[31] Wu CY, Qian PJ, Yu NJ. The fusion-driven hybrid system and its material selection. J Nucl Mater 2002;307–11(2):1629–36.

[32] Jackson RT, Liu H, Patrikalakis MN, Sachs ME, Cima JM. Modeling and designing functionally graded material components for fabrication with local composition control. Mater Des 1999;20 (2–3):63–75.

[33] Huda Z, Edi P. Materials selection in design of structures and engines of supersonic aircrafts: a review. Mater Des 2013;46:552–60.

[34] Chmielewski M, Pietrzak K. Metal-ceramic functionally graded materials – Manufacturing, characterization, application. Bull Polish Acad Sci Tech Sci 2016;64(1):151–60.

[35] CES S. Visualizing and communicate materials properties, Materials property charts: sample for Ashby charts. GRANTA Mater Intell Sample pro, 2017.

[36] Farag MM.Selection substitution sources of information. 2015.

[37] Ljungberg YL. Materials selection and design for development of sustainable products. Mater Des 2007;28(2):466–79.

[38] Fayazbakhsh K, Abedian A, Manshadi DB, Khabbaz SR. Introducing a novel method for materials selection in mechanical design using Z-transformation in statistics for normalization of material properties. Mater Des 2009;30(10):4396–404.

[39] Giudice F, La Rosa G, Risitano A. Materials selection in the life-cycle design process: a method to integrate mechanical and environmental performances in optimal choice. Mater Des 2005;26 (1):9–20.

[40] Thakker A, Jarvis J, Buggy M, Sahed A. A novel approach to materials selection strategy case study: Wave energy extraction impulse turbine blade. Mater Des 2008;29(10):1973–80.

[41] De Boer L, Labro E, Morlacchi P. A review of methods supporting supplier selection. Eur J Purch Supply Manag 2001;7(2):75–89.

[42] Ashby FM, Bréchet YJ, Cebon D, Salvo L. Selection strategies for materials and processes. Mater Des 2004;25(1):51–67.

[43] Mahamood MR, Member ET, Shukla M, Pityana S.Functionally graded material: an overview. World Congr Eng 2012;III:2–6.

[44] Rabboh AS, Bondok EN, Mahmoud ST, El Kholy HI. The effect of functionally graded materials into the sandwich beam dynamic performance. Materials Sciences and Applications. 2013;2013 (November):751–60.

[45] Ray CM, Sachade MH. Finite element analysis of smart functionally graded plates. Int J Solids Struct 2006;43(18–19):5468–84.

[46] Jamian S, Ayob NS, Abidin MR, Muhd Nor NH. Fabrication of functionally graded natural fibre/ epoxy cylinder using centrifugal casting method. ARPN J Eng Appl Sci 2016;11(4):2327–31.

[47] Das S, Sarangi KS. Static analysis of functionally graded composite beams. IOP Conf Ser Mater Sci Eng 2016;149:12138.

[48] Jackson TR. Analysis of functionally graded material object representation methods. Diss. Massachusetts Institute of Technology, 2000.

[49] Tambyrajah D. Indulge & Explore Natural Fiber Composites. An invitation to product designers. 2015.

[50] Tan T, Rahbar N, Allameh SM, Kwofie S, Dissmore D, Ghavami K, Soboyejo WO. Mechanical properties of functionally graded hierarchical bamboo structures. Acta biomaterialia 7.10 (2011):3796–3803.

[51] Ghavami K, Rodrigues CS, Paciornic S. Bamboo functionally graded composite material. Asian J Civ Eng 2003;4(January):1–10.

[52] Pereira PHF, Rosa MDF, Cioffi MOH, Benini KCCDC, Milanese AC, Voorwald HJC, Mulinari DR. Vegetal fibers in polymeric composites: a review. Polímeros 25.1 (2015):9–22.

[53] Mohammed L, Ansari MNM, Pua G, Jawaid M, Islam SM. A review on natural fiber reinforced polymer composite and its applications. Int J Polym Sci 2015;2015:1–15.

Savita K. Subramaniam, Vivek Kumar Gaba,
and Shubhankar Bhowmick

6 Temperature distribution in functionally graded longitudinal fins of varying geometry

Abstract: In the present chapter, performance of functionally graded longitudinal fins having varying geometry is reported. A comparison of the performance of the parabolic longitudinal fins having fixed aspect ratio with longitudinal transition in thermal conductivity approximated by means of a power series is reported by solving the second order governing differential equation. The fin is insulated at the tip with a predefined temperature at the base. A parametric study is then carried out by controlling the geometry parameters and grading parameters. The solution has been obtained using *bvp4c* subroutine, a technique of solving boundary value problems using continuous piecewise polynomial that adjusts the boundary conditions and is validated with benchmark results. Further, the results have been reported for combinations of geometry and grading parameters and are presented in graphical and tabular forms. The results give a substantial insight into the behavior of longitudinal fins and can be used as design data.

Keywords: functionally graded, power series, parabolic longitudinal fins

6.1 Introduction

Since long ago, fins are used as surfaces extended for increasing heat transfer rate through heat exchangers. Heat dissipation through fins involves a primary surface and a surrounding fluid [1]. The use of fins increases the surface area and thus helps to increase the heat transfer rate [2]. Myriad number of contributions have been made till date to analyze the design of fin geometry based on various parameters and thus to achieve optimum design criteria for different situations. One of the most significant assumptions, which was made to simplify the analysis of fins, was the assumption of negligible temperature variation along the fin cross section. This greatly reduced the mathematical equation from partial to ordinary, which led to validate results from well-established analytical solutions for a number of cases. But actually these thermophysical properties vary with temperature and materials [3, 4]. Kulkarni and Joglekar [5] proposed a solution of nonlinear differential of the temperature distribution in longitudinal fins with temperature-dependent thermal conductivity based on residue minimization. To evaluate temperature distribution within the fins,

https://doi.org/10.1515/9783110545104-006

Domairry and Fazeli [6] solved nonlinear differential equation for a straight fin by homotopy analysis method. The effects of various parameters, such as porosity, Darcy number, and Lewis number in a wet circular porous fin, on the fin efficiency have also been developed [7].

Further, the heat transfer coefficient for conduction may vary in accordance with the temperature difference between the surrounding and the surface, and spatial coordinate. In addition to this, the convective heat transfer coefficient depends on the local temperature difference (power law dependence) and the convective process. The power law dependence can be found in Kraus et al. [2]. The effects of radiation come into play at high temperatures. In industrial applications, where engineering processes occur at high temperature, radiation effects are to be considered [8–10]. Assuming convection and radiation sink temperature, Torabi et al. [11] reported Differential Transform Method (DTM)-based solution of energy equation in moving fins having variable thermal conductivity. The efficiency and performance of convective–radiative longitudinal varying geometry fins is reported by Torabi and Zang [12]. Singla and Das [13] described the parametric effects in a moving fin using inverse prediction. For temperature-dependent longitudinal fins of varying thickness, having internal heat generation, Sobhan et al. [14] proposed the solution based on transient analysis. Mokheimer [15] investigated the effect of locally variable heat transfer coefficient on the performance of convective fins of varying profiles. Aziz [16] reported the use of Bessel functions to calculate the optimum fin dimensions of longitudinal, annular fins and spines of varying profiles. Ullmann and Kalman [17] reported the optimum dimensions and efficiency of parabolic and hyperbolic radial fins.

Meanwhile, critical development resulted in parallel regarding functionally graded materials [18, 19]. Aziz and Fang [20] reported the behavior of functionally graded rectangular radial fin having power law conductivity variation. The results obtained in terms of Airy wave functions, power functions, as well as Bessel functions were compared with the results from spatially averaged thermal conductivity model. In a recent work, Gaba et al. [21, 22] reported the performance of functionally graded rotating annular fins [21] and constant weight FG annular fins of parabolic thickness variation respectively.

Present work reports the effect of functional grading on thermal performance in a longitudinal fin of varying thickness. Also, the influence of geometry on the performance of functionally graded longitudinal fin is studied.

6.2 Mathematical formulation

Longitudinal (straight) fin of constant width w is shown in Figure 6.1. The fin is conductive–convective, and the effect of radiation is neglected. The convection coefficient, h, is assumed constant throughout the fin surface. Different fin

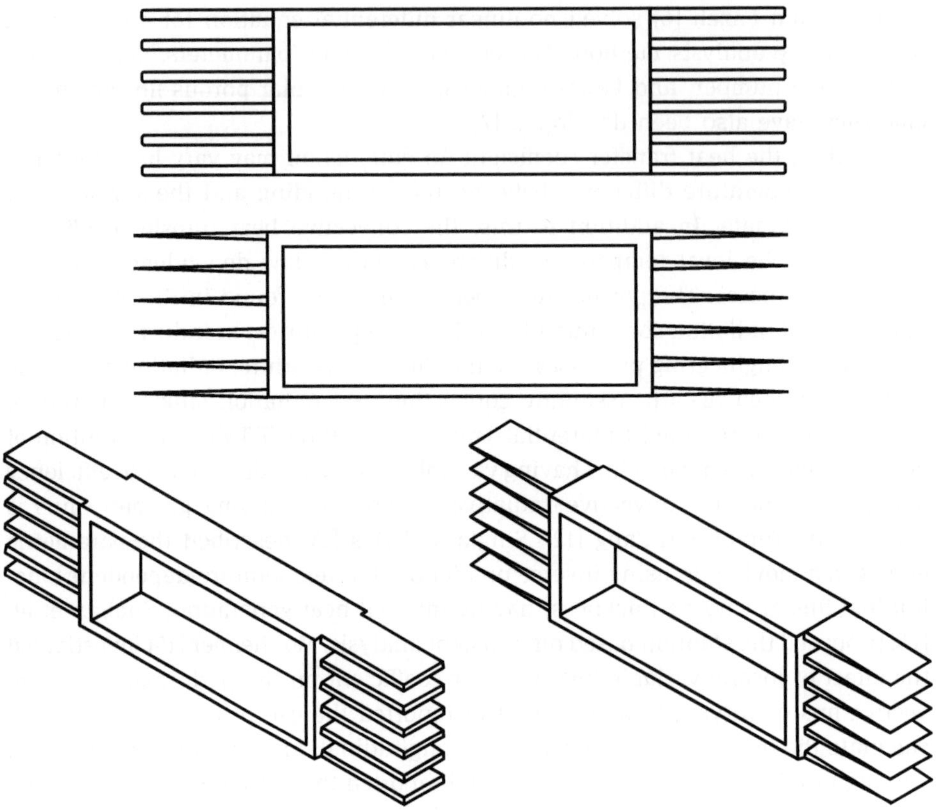

Figure 6.1: Sectional view of a longitudinal fin with constant area.

parameters are cross section Ac, perimeter P, and width w. The thickness of fin is presented by δ. Base of the fin is subjected to a constant temperature, while the free end is insulated. The heat sink temperature is taken as T_a. The conductivity K is taken to be a power function of x as

$$K = a\ (1-X)^b \tag{6.1}$$

where a is the conductivity coefficient at b = 0 and a has the dimension of W/m^2. Based upon these conditions two cases are taken, first where the width of the fin is constant and second where width is varying across the fin length.

The longitudinal axis x is measured from the base of the fin along the fin length:

$$\frac{d^2T}{dx^2} + \left(\frac{b * x^{(b-1)}}{(x)^b}\right)\frac{dT}{dx} = \left(\frac{h \times 2 \times (\mathrm{w} + \delta_0)}{A \times a(x)^b}\right)(T - Ta) \tag{6.2}$$

Equation (2) is solved using the two boundary conditions stated above and treated mathematically as follows:

$$x = 0;\ T = T_b$$
$$x = L;\ \frac{dT}{dx} = 0 \tag{6.3}$$

The equations are reduced in normalized form, as

$$\theta = \frac{T}{T_b};\ \theta_a = \frac{T_a}{T_b};\ \theta_b = 1;\ X = \frac{x}{L} \tag{6.4}$$

The equation of a functionally graded fin with constant thickness in dimensionless form is as follows:

$$\frac{d^2\theta}{dX^2} + \left(\frac{b * L^b * X^{(b-1)}}{(X * L)^b}\right)\frac{d\theta}{dX} = \left(\frac{L^2 * h * 2 * (\mathrm{w} + \delta_0)}{(A * a * (X * L)^b)}\right)(\theta - \theta a) \tag{6.5}$$

For a functionally graded varying geometry fin (Figure 6.2) with thermal conductivity *K*, obtained from Eq. (1), and thickness, being a parabolic function of the length coordinate is given as

$$\delta = \delta_0(1 - mX^n) \tag{6.6}$$

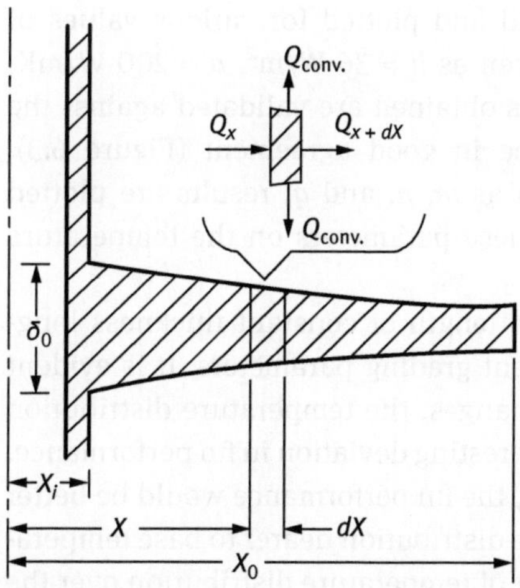

Figure 6.2: Sectional view of a parabolic longitudinal fin.

where m and n are dimensionless constants and δ_o is the thickness of the fin at base. With these variations, new heat transfer relation in normalized form is obtained as

$$\frac{d^2\theta}{dX^2}+\left(\frac{-b}{(1-X)}+\frac{-mn(X)^{n-1}}{1-mX^n}\right)\frac{d\theta}{dX}-\frac{2h\times L^2(\delta+w)}{kw\delta o(1-mX^n)}\theta=0 \tag{6.7}$$

The above equation with variation of thermophysical characteristics is solved in MATLAB using *bvp4c* routine. An established way of solving boundary value problems is to select a spline with unknown coefficients that adjusts the boundary conditions. The unknown coefficients are determined by suitable collocation of the algebraic equations at quadrature points. Code *bvp4c* [22] is, hence, based on an adaptive finite difference algorithm with three-stage (Lobatto III-a) collocation formula for solving the spline in mesh subinterval $[x_i, x_{i+1}]$. The important requirements are attainment of the boundary conditions and collocation of the residue at endpoints and midpoint of the subinterval. The approximate solution is obtained by residue minimization over all mesh subintervals using Lobatto 5-point quadrature formula.

For validating eqs. (6.4) and (6.7) obtained for rectangular and parabolic thickness variation, respectively, the values of m, n, and b could be substituted with zero and the resulting second-order ODE is solved using *bvp4c* subroutine in MATLAB.

6.3 Results and discussions

The normalized temperature, θ, being a function of nondimensional variables, depends on geometry parameters n and m, and on a, q, and R, due to functional grading along the axis. Thus, considering the length coordinate, x, being the only independent variable, eq. (6.7) is solved and plotted for various values of m, n, a, and b. The system properties are taken as h = 36 W/m^2, a = 200 W/mK, δ_a = 0.02 m, and R_1 = 0.2 m. The results thus obtained are validated against the benchmark results [23] and are found to be in good agreement (Figure 6.3). Further upon varying other parameters such as m, n, and q, results are plotted to interpret the relation between effects of these parameters on the temperature profile (Figures 6.4–7).

The temperature distribution over the fin length of constant thickness longitudinal fin is plotted in Figure 6.4 for different grading parameters. It is evident from the figure that, as grading parameter changes, the temperature distribution also changes significantly. This leads to an interesting deviation in fin performance. It is evident that for grading parameter b = −2, the fin performance would be better than the remaining as the fin with temperature distribution nearer to base temperature yields maximum efficiency. A similar plot of temperature distribution over the fin length for parabolic fins with functional graded material is shown in Figure 6.5.

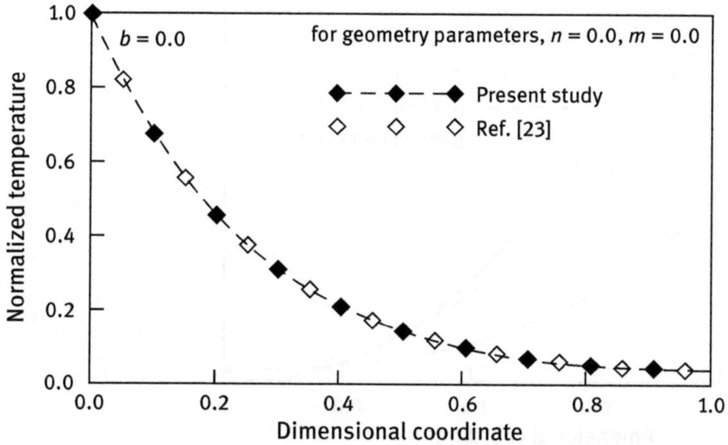

Figure 6.3: Validation of present work with general equation of fin.

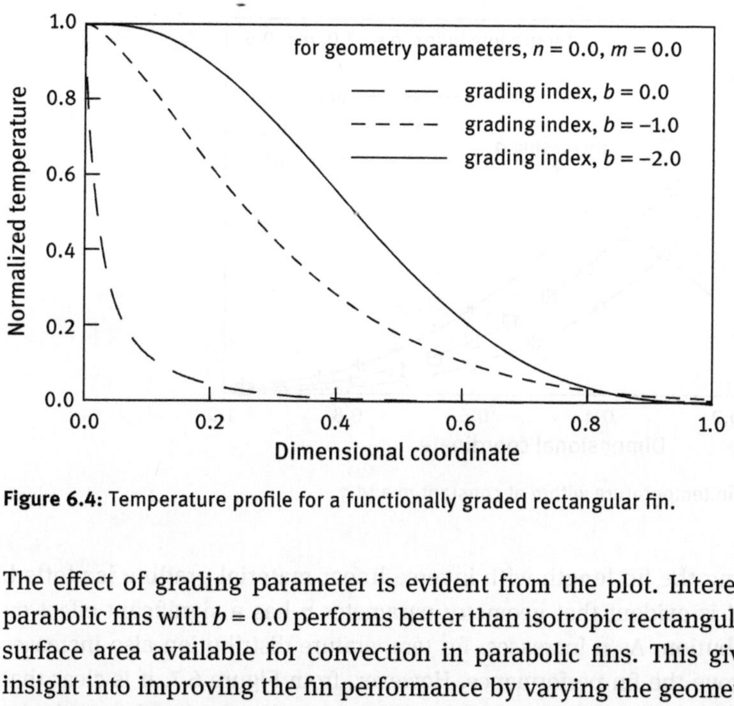

Figure 6.4: Temperature profile for a functionally graded rectangular fin.

The effect of grading parameter is evident from the plot. Interestingly, isotropic parabolic fins with $b = 0.0$ performs better than isotropic rectangular fin due to more surface area available for convection in parabolic fins. This gives an important insight into improving the fin performance by varying the geometry. However, the investigation of the effect of grading parameter on parabolic fins reveal that the fin performance does not substantially change with that of rectangular fins at same grading. This shows that fin performance could be enhanced by either varying the geometry or varying the conductivity. Varying both at a time could be a difficult and costly proposition to fabricate and may not yield results at par with the investment. In Figures 6.6 and 6.7, the effect of geometry parameters on temperature profile of

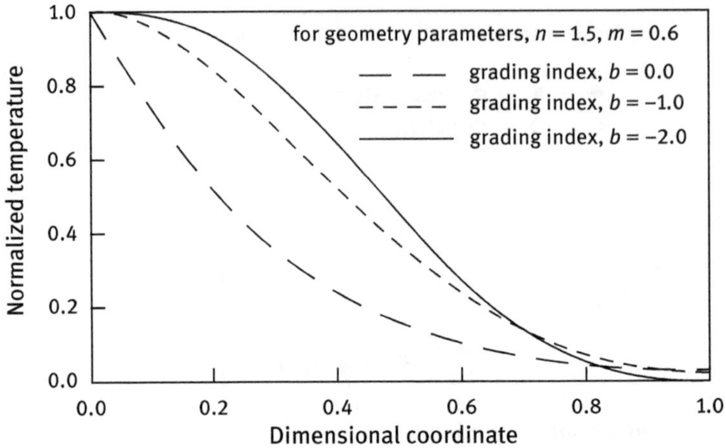

Figure 6.5: Temperature profile for a functionally graded parabolic fin.

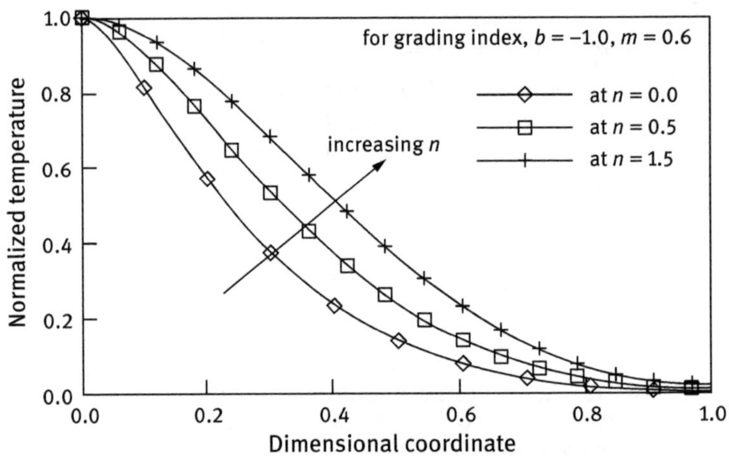

Figure 6.6: Variation in temperature with n at constant m and q.

parabolic fins along the fin length with inverse linear material grading is plotted. From Figure 6.6 it is evident that geometry parameter n has a significant effect on temperature distribution. As n increases, fin temperature distribution also increases and tends to improve the fin performance. However, from Figure 6.7, it is clear that geometry parameter m has negligible effect on temperature distribution. Interestingly, as effect of m on the temperature distribution is insignificant, hence for varying geometry fins, the profile selected could be a linearly tapered one (i.e., m = 1.0) being easy to fabricate. It should be realized that varying m changes the profile while the variation of n affects the thickness over the fin length.

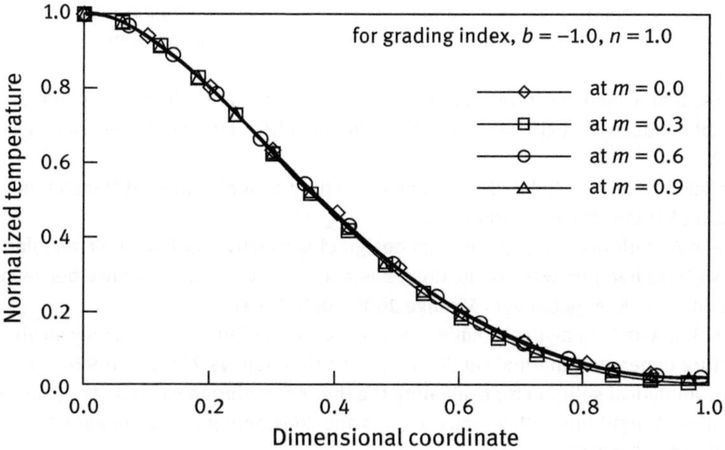

Figure 6.7: Variation of temperature with m at constant n and q.

6.4 Conclusion

The performance of longitudinal constant thickness and parabolic fins is reported. The study is presented by varying geometry parameters n and m. The temperature distribution of longitudinal fins along the fin length is reported at different grading parameter b. The effect of grading parameter on fin temperature is established for rectangular and parabolic fins. The effect of geometry parameter n is also presented. The chapter points out toward an interesting conclusion that the fin performance could be improved by either varying geometry or by varying the conductivity of fin material using negative grading parameter.

References

[1] Kern QD, Kraus AD. Extended surface heat transfer. New York: McGraw-Hill, 1972.
[2] Kraus AD, Aziz A, Welty JR. Extended surface heat transfer. New York: John Wiley, 2002.
[3] Aziz A, Benzies JY. Application of perturbation techniques to heat-transfer problems with variable thermal properties. Int J Heat Mass Transfer 1976;19:271–6.
[4] Ganji DD, Rahimi M, Rahgoshay M, Jafari M. Analytical and numerical investigation of conductive, convective and radiative straight fins. Heat Trans Asian Res 2011;40:233–45.
[5] Kulkarni DB, Joglekar MM. Residue minimization technique to analyze the efficiency of convective straight fins having temperature-dependent thermal conductivity. Appl Math Comput 2009;215(6):2184–91.
[6] Domairry G, Fazeli M. Homotopy analysis method to determine the fin efficiency of convective straight fins with temperature-dependent thermal conductivity. Commun Nonlinear Sci Numer Simul 2009;14:489–99.

[7] Hatami M, Ganji DD. Investigation of refrigeration efficiency for fully wet circular porous fins with variable sections by combined heat and mass transfer analysis. Int J Refrig 2014;40:140–51.

[8] Bouaziz MN, Aziz A. Simple and accurate solution for convective–radiative fin with temperature dependent thermal conductivity using double optimal linearization. Energy Convers Manage 2010;51:2776–82.

[9] Bouaziz MN, Rechak S, Hanini S, Bal Y, Bal K. Numerical study of nonlinear heat transfer in longitudinal fins. Int J Therm Sci 2001;40:843–57.

[10] Aziz A, Beers-Green A. Performance and optimum design of convective–radiative rectangular fin with convective base heating, wall conduction resistance, and contact resistance between the wall and the fin base. Energy Convers Manage 2009;50:2622–31.

[11] Torabi M, Yaghoobi H, Aziz A. Analytical solution for convective–radiative continuously moving fin with temperature dependent thermal conductivity. Int J Thermophys 2012;33(5):924–41.

[12] Torabi M, Zang Q. Analytical solution for evaluating the thermal performance and efficiency of convective–radiative straight fins with various profiles and considering all non-linearities. Energy Convers Manage 2013;66:199–210.

[13] Singla R, Das R. Application of decomposition method and inverse prediction of parameters in a moving fin. Energy Convers Manage 2014;84:268–281.

[14] Mosayebidorcheh S, Farzinpoor M, Ganji DD. Transient thermal analysis of longitudinal fins with internal heat generation considering temperature-dependent properties and different fin profiles. Energy Convers Manage 2014;86:365–70.

[15] Mokheimer EM. Heat transfer from extended surfaces subject to variable heat transfer coefficient. Heat Mass Transfer 2003;39(2):131–8.

[16] Aziz, A. Optimum Dimensions of extended surfaces operating in a convective environment. Appl Mech Rev 1992;45(5):155–73.

[17] Ullmann A, Kalman H. Efficiency and optimized dimensions of annular fins of different cross-section shapes. Int J Heat Mass Transfer 1989;32(6):1105–10.

[18] Koizumi M, Niino M. Overview of FGM research in Japan. MRS Bulletin 1995;20(1):19–21. doi:10.1557/S0883769400048867

[19] Cannillo V, Lusvarghi L, Manfredini T, Montorsi M, Siligardi C, Sola A. Functionally graded materials: prevision of properties and performances, OOF Workshop–24, Dipartimento di Ingegneria dei Materiali e dell'Ambiente, University of Modena and Reggio Emilia – Italy, 25 August 2006.

[20] Aziz A, Fang T. Thermal analysis of an annular fin with (a) simultaneously imposed base temperature and base heat flux and (b) fixed base and tip temperatures. Energy Convers Manage 2011;52(7):2467–78. DOI:https://doi.org/10.1016/j.enconman.2011.02.004

[21] Gaba VK, Tiwari AK, Bhowmick S. A Parametric study of functionally graded rotating annular fin. Procedia Eng 2015;127:126–32.

[22] Shampine LF, Reichelt MW, Kierzenka J. Solving boundary value problems for ordinary differential equations in MATLAB with bvp4c. 2006. Available from <ftp://ftp.mathworks.com/pub/doc/papers/bvp/>.

[23] Çengel YA, Ghajar AJ. Heat and mass transfer: fundamentals & applications, 5th ed. New York: McGraw Hill Education, 2015.

Ta Duy Hien

7 Analytic approach for transient response of functionally graded rectangular plates including the higher-order shear deformation effects

Abstract: This chapter deals with the analytic solution for transient response of functionally graded (FG) rectangular plates under transverse loadings. The governing equations account for higher-order shear deformation in FG plates. The material properties of the plates are assumed to vary continuously in the thickness direction according to the power-law form. Analytical solution based on state variable method is presented for both free and forced vibration of simply supported FG rectangular plates. The results obtained using analytical approach are compared with those obtained by finite element method. The effects of power-law exponent index on the behavior of FG plates are also studied.

Keywords: state variable, dynamic response, functionally graded plate, higher-order plate theory

7.1 Introduction

Functionally graded (FG) materials are a new type of composite structures that attracts lots of interest for research and engineering design. The FG material is made of two or more material components. They are manufactured to vary continuously along the thickness direction. Normally, FG materials are manufactured by a compositional gradient from a ceramic to a metal. Besides artificial material, FG materials exist naturally such as bamboo [1, 2].

The popular shear deformation plate theories used for composite plates and FG plates are the classical plate theory (CPT), the first-order shear deformation theory (FSDT), and higher-order shear deformation theory (HSDT). The CPT ignores the transverse shear deformation and it is appropriate for thin structures. The FSDT is a simple shear deformation theory and it can be used for both thin and thick plates. However, it usually uses a shear correction factor in which their values are quite dispersed through each problem. The HSDT is not necessary a shear correction factor, but on the other hand, governing equations are more complex than those from the FSDT. However, the HSDT gives more accurate transverse shear stresses. There are many HSDTs proposed [3–6].

There are many works in the literature on free vibration of the plates. Investigating free vibration of composite plates uses analytical method accounting

https://doi.org/10.1515/9783110545104-007

FSDT [7–9]), HSDTs [10, 11], and various shear deformation theories [12]. Besides the analytical method, numerical methods can be used as finite element method [13–15]), element-free Galerkin method [16], and isogeometric analysis [17].

The transient response of plates has attracted the attention of lots of researchers. The transient response of plate can be determined by the analytic method [18–20], the finite element method [21–23], finite difference method [24]. The analytical method has many restrictions to be applied to the special plates such as rectangular plate and circular plate with simple boundary conditions. On the other hand, almost all dynamics of plate problems can use numerical methods for finding responses.

In relevance to FG plates, the free vibration of FG plates was investigated using analytical method [25–27], finite element methods [28], isogeometric analysis [29], and element-free K_p-Ritz method [30]. In case of the transient response of FG plates, few reports have been made since the appearance of FG plates. Yang et al. [31] investigated the dynamic response of initially stressed FG rectangular plate using modal superposition method based on combining one-dimensional differential quadrature approximation and the Galerkin procedure. Yang et al. [32] studied the nonlinear transient response of FG plates by combining perturbation approach, Galerkin technique, and Runge–Kutta method. Wen et al. [33] solved the three-dimensional elastic dynamic problem for the simply supported FG plates using radial basis function method. Reddy [34] used the finite element method to determine the transient response of FG plates accounting third shear deformation and the von Kármânan-type geometric nonlinearity.

In this work, the state variable method is used to find analytical solutions for the transient response of FG rectangular plate. The simply supported FG rectangular plate subjected to distributed dynamic loads is investigated with time functions as a step, sine, and triangular pulse. The dynamic responses predicted by the analytical approach using HSDT are verified with finite element method by using commercial software.

7.2 Governing equations and solution procedures

7.2.1 The plate model

Consider an FG plate made of ceramic and metal, which has the geometry of the plate and coordinate system (x,y,z) is shown in Figure 7.1.

Suppose that Young's modulus E and mass density ρ of an FG plate varies through the plate thickness according to the formulas:

$$\begin{aligned} \rho(z) &= \rho_m + (\rho_c - \rho_m)\left(\frac{z}{h} + \frac{1}{2}\right)^k \\ E(z) &= E_m + (E_c - E_m)\left(\frac{z}{h} + \frac{1}{2}\right)^k \end{aligned} \tag{7.1}$$

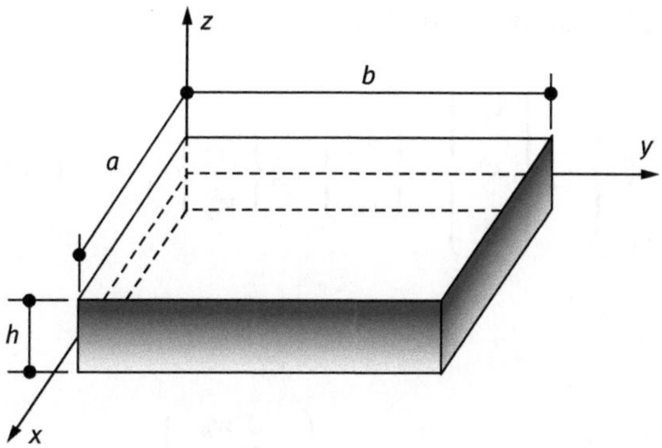

Figure 7.1: Model of a rectangular plate.

where k is the power-law index of the volume fraction, and the subscripts c and m represent the ceramic and metallic component, respectively. The linear constitutive equation of the FG plate is given as follows:

$$\begin{Bmatrix} \sigma_x \\ \sigma_y \\ \sigma_{yz} \\ \sigma_{xz} \\ \sigma_{xy} \end{Bmatrix} = \begin{bmatrix} \tilde{Q}_{11} & \tilde{Q}_{12} & 0 & 0 & 0 \\ \tilde{Q}_{12} & \tilde{Q}_{22} & 0 & 0 & 0 \\ 0 & 0 & \tilde{Q}_{44} & 0 & 0 \\ 0 & 0 & 0 & \tilde{Q}_{55} & 0 \\ 0 & 0 & 0 & 0 & \tilde{Q}_{66} \end{bmatrix} \begin{Bmatrix} \varepsilon_x \\ \varepsilon_y \\ \gamma_{yz} \\ \gamma_{xz} \\ \gamma_{xy} \end{Bmatrix} \tag{7.2}$$

where

$$\tilde{Q}_{11} = \frac{E}{1-\nu^2}, \quad \tilde{Q}_{22} = \frac{E}{1-\nu^2}, \quad \tilde{Q}_{12} = \frac{\nu E}{1-\nu^2}$$

$$\tilde{Q}_{44} = G_{23}, \quad \tilde{Q}_{55} = G_{13}, \quad \tilde{Q}_{66} = G_{12}$$

with the Poisson's ratio ν is supposed to be constant.

Shimpi and Patel [5] proposed the HSDT for the orthotropic plate. Extending this HSDT for the FG plate, the displacement fields can be written as

$$\begin{aligned} U(x,y,z,t) &= u(x,y,z,t) - \frac{\partial w_b}{\partial x} z + \frac{\partial w_s}{\partial x}\left[\frac{1}{4}z - \frac{5}{3}z\left(\frac{z}{h}\right)^2\right] \\ V(x,y,z,t) &= v(x,y,z,t) - \frac{\partial w_b}{\partial y} z + \frac{\partial w_s}{\partial y}\left[\frac{1}{4}z - \frac{5}{3}z\left(\frac{z}{h}\right)^2\right] \\ W(x,y,z,t)x &= w_b(x,y,t) + w_s(x,y,t) \end{aligned} \tag{7.3}$$

where (U,V,W) are the displacements at coordinate (x, y, z) and (u, v, w_b, w_s) are the displacements of the midplane of the plate.

The linear strains can be obtained by deriving eq. (7.3) as

$$\begin{Bmatrix} \varepsilon_x \\ \varepsilon_y \\ \gamma_{xy} \end{Bmatrix} = \begin{Bmatrix} \varepsilon_x^0 \\ \varepsilon_y^0 \\ \gamma_{xy}^0 \end{Bmatrix} + z \begin{Bmatrix} \kappa_x^b \\ \kappa_y^b \\ \kappa_{xy}^b \end{Bmatrix} + \tilde{f} \begin{Bmatrix} \kappa_x^s \\ \kappa_y^s \\ \kappa_{xy}^s \end{Bmatrix}, \quad \begin{Bmatrix} \gamma_{yz} \\ \gamma_{xz} \end{Bmatrix} = \hat{g} \begin{Bmatrix} \gamma_{yz}^s \\ \gamma_{xz}^s \end{Bmatrix} \tag{7.4}$$

where

$$\{\varepsilon^0\} = \begin{Bmatrix} \varepsilon_x^0 \\ \varepsilon_y^0 \\ \gamma_{xy}^0 \end{Bmatrix} = \begin{Bmatrix} \dfrac{\partial u}{\partial x} \\ \dfrac{\partial v}{\partial y} \\ \dfrac{\partial u}{\partial y} + \dfrac{\partial v}{\partial x} \end{Bmatrix}, \quad \{\kappa^b\} = \begin{Bmatrix} \kappa_x^b \\ \kappa_y^b \\ \kappa_{xy}^b \end{Bmatrix} = \begin{Bmatrix} -\dfrac{\partial^2 w_b}{\partial x^2} \\ -\dfrac{\partial^2 w_b}{\partial y^2} \\ -2\dfrac{\partial^2 w_b}{\partial x \partial y} \end{Bmatrix},$$

$$\{\kappa^s\} = \begin{Bmatrix} \kappa_x^s \\ \kappa_y^s \\ \kappa_{xy}^s \end{Bmatrix} = \begin{Bmatrix} -\dfrac{\partial^2 w_s}{\partial x^2} \\ -\dfrac{\partial^2 w_s}{\partial y^2} \\ -2\dfrac{\partial^2 w_s}{\partial x \partial y} \end{Bmatrix}, \tag{7.5}$$

$$\begin{Bmatrix} \gamma_{yz} \\ \gamma_{xz} \end{Bmatrix} = \begin{Bmatrix} \dfrac{\partial w_s}{\partial y} \\ \dfrac{\partial w_s}{\partial x} \end{Bmatrix}, \quad \hat{g}(z) = 5\left[\frac{1}{4} - \left(\frac{z}{h}\right)^2\right], \quad \tilde{f}(z) = -\frac{1}{4}z + \frac{5}{3}z\left(\frac{z}{h}\right)^2$$

7.2.2 Governing equations

The differential equations of the behavior of FG plate will be derived using the energy principle. The first step is to determine the energy formulation of structures.

The kinetic energy of FG plate can be written as

$$T = \frac{1}{2}\int_{\Omega} \left\{ (\dot{w}_b + \dot{w}_s)^2 + \left(\dot{u} - z\frac{\partial \dot{w}_b}{\partial x} - \tilde{f}\frac{\partial \dot{w}_s}{\partial x}\right)^2 + \left(\dot{v} - z\frac{\partial \dot{w}_b}{\partial y} - \tilde{f}\frac{\partial \dot{w}_s}{\partial y}\right)^2 \right\} \rho(z) d\Omega \tag{7.6}$$

where $\rho(z)$ is the mass density in the plate.

The potential energy of the external load is

$$V = -\int_A \hat{p}(w_b + w_s)dxdy \tag{7.7}$$

where $\hat{p}$ is the transverse load on the plate.

The strain energy stored in the FG plate is as follows:

$$\Pi = \frac{1}{2}\int_\Omega \left(\sigma_x\varepsilon_x + \sigma_y\varepsilon_y + \sigma_{xy}\gamma_{xy} + \sigma_{xz}\gamma_{xz} + \sigma_{yz}\gamma_{yz}\right)d\Omega \tag{7.8}$$

Substituting eqs (7.2) and (7.4) into eq. (7.8), and then integrate through the thickness of the plate, the formula for the strain energy of the plate is obtained as follows:

$$\begin{aligned}\Pi = \tfrac{1}{2}\int_A [&N_x\varepsilon_x^0 + N_y\varepsilon_y^0 + N_{xy}\gamma_{xy}^0 + Q_x\gamma_{xz}^s + Q_y\gamma_{yz}^s + M_x^b\kappa_x^b + M_y^b\kappa_y^b + M_{xy}^b\kappa_{xy}^b \\ &+ M_x^s\kappa_x^s + M_y^s\kappa_y^s + M_{xy}^s\kappa_{xy}^s]dxdy\end{aligned} \tag{7.9}$$

where

$$\begin{cases} N_x = \int_{-h/2}^{h/2}\sigma_x dz; & N_y = \int_{-h/2}^{h/2}\sigma_y dz; & N_{xy} = \int_{-h/2}^{h/2}\sigma_{xy} dz \\ M_x^b = \int_{-h/2}^{h/2} z\sigma_x dz; & M_y^b = \int_{-h/2}^{h/2} z\sigma_y dz; & M_{xy}^b = \int_{-h/2}^{h/2} z\sigma_{xy} dz \\ M_x^s = \int_{-h/2}^{h/2}\tilde{f}\sigma_x dz; & M_y^s = \int_{-h/2}^{h/2}\tilde{f}\sigma_y dz; & M_{xy}^s = \int_{-h/2}^{h/2}\tilde{f}\sigma_{xy} dz \\ Q_x = \int_{-h/2}^{h/2}\hat{g}\sigma_{xz} dz; & Q_y = \int_{-h/2}^{h/2}\hat{g}\sigma_{yz} dz & \end{cases} \tag{7.10}$$

and notation block matrices:

$$\begin{cases} \{N\} = \{N_x \quad N_y \quad N_{xy}\}^T \\ \{M^b\} = \{M_x^b \quad M_y^b \quad M_{xy}^b\}^T \\ \{M^s\} = \{M_x^s \quad M_y^s \quad M_{xy}^s\}^T \end{cases}. \tag{7.11}$$

We can obtain the force and moment resultants by using the stress–strain relations as follows:

$$\begin{gathered}\begin{Bmatrix}\{N\}\\ \{M^b\}\\ \{M^s\}\end{Bmatrix} = \begin{bmatrix}[A] & [B] & [B^s]\\ [B] & [D] & [D^s]\\ [B^s] & [D^s] & [H^s]\end{bmatrix}\begin{Bmatrix}\{\varepsilon^0\}\\ \{\kappa^b\}\\ \{\kappa^s\}\end{Bmatrix} \\ \begin{Bmatrix}Q_y\\ Q_x\end{Bmatrix} = \begin{bmatrix}A_{44}^s & 0\\ 0 & A_{55}^s\end{bmatrix}\begin{Bmatrix}\gamma_{yz}^s\\ \gamma_{xz}^s\end{Bmatrix}\end{gathered} \tag{7.12}$$

where

$$[A] = \begin{bmatrix} A_{11} & A_{12} & 0 \\ A_{12} & A_{22} & 0 \\ 0 & 0 & A_{66} \end{bmatrix}, [B] = \begin{bmatrix} B_{11} & B_{12} & 0 \\ B_{12} & B_{22} & 0 \\ 0 & 0 & B_{66} \end{bmatrix}$$
$$[D] = \begin{bmatrix} D_{11} & D_{12} & 0 \\ D_{12} & D_{22} & 0 \\ 0 & 0 & D_{66} \end{bmatrix}, [B^s] = \begin{bmatrix} B^s_{11} & B^s_{12} & 0 \\ B^s_{12} & X^s_{22} & 0 \\ 0 & 0 & B^s_{66} \end{bmatrix} \quad (7.13)$$
$$[D^s] = \begin{bmatrix} D^s_{11} & D^s_{12} & 0 \\ D^s_{12} & D^s_{22} & 0 \\ 0 & 0 & D^s_{66} \end{bmatrix}, [H^s] = \begin{bmatrix} H^s_{11} & H^s_{12} & 0 \\ H^s_{12} & H^s_{22} & 0 \\ 0 & 0 & H^s_{66} \end{bmatrix}$$

The plate stiffnesses A_{ij}, B_{ij}, etc., are given by

$$\begin{cases} A_{ij} = \int_{-h/2}^{h/2} \tilde{Q}_{ij} dz \\ B_{ij} = \int_{-h/2}^{h/2} z\tilde{Q}_{ij} dz \\ D_{ij} = \int_{-h/2}^{h/2} z^2\tilde{Q}_{ij} dz \\ B^s_{ij} = \int_{-h/2}^{h/2} \tilde{f}\tilde{Q}_{ij} dz \\ D^s_{ij} = \int_{-h/2}^{h/2} z\tilde{f}\tilde{Q}_{ij} dz \\ H^s_{ij} = \int_{-h/2}^{h/2} \tilde{f}^2\tilde{Q}_{ij} dz \end{cases} \quad \text{with index } i=1,2,6$$
$$A^s_{ij} = \int_{-h/2}^{h/2} \hat{g}\tilde{Q}_{ij} dz \qquad \text{with index } i=4,5$$

Using variational principles, we can obtain the differential equations of motion of the plate as follows:

$$\begin{aligned} \frac{\partial N_{xy}}{\partial y} &+ \frac{\partial N_x}{\partial x} = I_0\ddot{u} - I_1\frac{\partial \ddot{w}_b}{\partial x} - J_1\frac{\partial \ddot{w}_s}{\partial x} \\ \frac{\partial N_{xy}}{\partial x} &+ \frac{\partial N_y}{\partial y} = I_0\ddot{v} - I_1\frac{\partial \ddot{w}_b}{\partial y} - J_1\frac{\partial \ddot{w}_s}{\partial y} \\ \frac{\partial^2 M^b_x}{\partial x^2} &+ 2\frac{\partial^2 M^b_{xy}}{\partial x\partial y} + \frac{\partial^2 M^b_y}{\partial y^2} + q = I_0(\ddot{w}_b + \ddot{w}_s) + I_1\left(\frac{\partial \ddot{u}}{\partial x} + \frac{\partial \ddot{v}}{\partial y}\right) - I_2\nabla^2\ddot{w}_b - J_2\nabla^2\ddot{w}_s \\ \frac{\partial^2 M^s_x}{\partial x^2} &+ 2\frac{\partial^2 M^s_{xy}}{\partial x\partial y} + \frac{\partial^2 M^s_y}{\partial y^2} + \frac{\partial Q_{xz}}{\partial x} + \frac{\partial Q_{yz}}{\partial y} + \hat{p} = I_0(\ddot{w}_b + \ddot{w}_s) + J_1\left(\frac{\partial \ddot{u}}{\partial x} + \frac{\partial \ddot{v}}{\partial y}\right) \\ &- J_2\nabla^2\ddot{w}_b - \hat{J}_1\nabla^2\ddot{w}_b \end{aligned} \quad (7.14)$$

where $q(x,y,t)$ is the transverse load, (N_i, Q_i, M_i^b, M_i^s) are the stress resultants and the inertias $\left(I_i, J_i, \hat{J}\right)$ are defined by

$$I_i = \int_{-h/2}^{h/2} \rho(z) z^i dz, \quad (i = 0, 1, 2, 3, 4, 6)$$

$$J_i = -\frac{I_i}{4} + \frac{I_{i+2}}{3h^2}, \quad (i = 1, 2)$$

$$\hat{J} = \frac{I_2}{16} - \frac{5I_4}{6h^2} + \frac{25I_6}{9h^2}$$

7.2.3 Solution procedures

The state variable approach can be used to solve the differential equations of motion for complicated structures. The state variable method mentioned by Dorf and Bishop [35] will be used to find the dynamic response of simply supported FG rectangular plates in eq. (7.14). The boundary conditions are simply supported as follows:

$$\begin{cases} v = w_b = w_s = 0, \quad \dfrac{\partial w_b}{\partial y} = \dfrac{\partial w_s}{\partial y} = 0 \\ N_x = 0, \quad M_x^b = M_x^s = 0 \end{cases} \quad \text{at } x = 0, a$$
$$\begin{cases} u = w_b = w_s = 0, \quad \dfrac{\partial w_b}{\partial x} = \dfrac{\partial w_s}{\partial x} = 0 \\ N_y = 0, \quad M_y^b = M_y^s = 0 \end{cases} \quad \text{at } y = 0, b \tag{7.15}$$

The Navier method used a sinusoidal function to satisfy all the boundary conditions in eq. (7.15). The displacement fields as series sinusoidal function are chosen as follows:

$$\begin{aligned} u(x,y,t) &= \sum_{n=1}^{\infty}\sum_{m=1}^{\infty} \{U_{mn}(t)\cos(\alpha x)\sin(\beta y)\} \\ v(x,y,t) &= \sum_{n=1}^{\infty}\sum_{m=1}^{\infty} \{V_{mn}(t)\sin(\alpha x)\cos(\beta y)\} \\ w_b(x,y,t) &= \sum_{n=1}^{\infty}\sum_{m=1}^{\infty} \{W_{bmn}(t)\sin(\alpha x)\sin(\beta y)\} \\ w_s(x,y,t) &= \sum_{n=1}^{\infty}\sum_{m=1}^{\infty} \{W_{smn}(t)\sin(\alpha x)\sin(\beta y)\} \end{aligned} \tag{7.16}$$

where $\alpha = \dfrac{m\pi}{a}$, $\beta = \dfrac{n\pi}{b}$

Substituting eq. (7.16) into eq. (7.14) and then applying the orthogonal property of the sinusoidal function, the differential equations of motion (7.14) are obtained as follows:

$$\begin{bmatrix} s_{11} & s_{12} & s_{13} & s_{14} \\ s_{12} & s_{22} & s_{23} & s_{24} \\ s_{13} & s_{23} & s_{33} & s_{34} \\ s_{14} & s_{24} & s_{34} & s_{44} \end{bmatrix} \begin{Bmatrix} U_{mn} \\ V_{mn} \\ W_{bmn} \\ W_{smn} \end{Bmatrix} + \begin{bmatrix} m_{11} & 0 & m_{13} & m_{14} \\ 0 & m_{22} & m_{23} & m_{24} \\ m_{13} & m_{23} & m_{33} & m_{34} \\ m_{14} & m_{24} & m_{43} & m_{44} \end{bmatrix} \begin{Bmatrix} \ddot{U}_{mn} \\ \ddot{V}_{mn} \\ \ddot{W}_{bmn} \\ \ddot{W}_{smn} \end{Bmatrix} = \begin{Bmatrix} 0 \\ 0 \\ F_{mn} \\ F_{mn} \end{Bmatrix} \tag{7.17}$$

where

$$\begin{aligned}
s_{11} &= A_{11}\alpha^2 + A_{66}\beta^2,\; s_{12} = (A_{12} + A_{66})\alpha\beta \\
s_{13} &= -\alpha\left[B_{11}\alpha^2 + (B_{12} + 2B_{66})\beta^2\right],\; s_{14} = B_{11}\alpha^2 + B_{66}\beta^2 \\
s_{22} &= A_{66}\alpha^2 + A_{22}\beta^2, s_{23} = -\beta\left[(B_{12} + 2B_{66})\alpha^2 + B_{22}\beta^2\right] \\
s_{24} &= -\beta\left[(B^s_{12} + 2B^s_{66})\alpha^2 + B^s_{22}\beta^2\right] \\
s_{33} &= D_{11}\alpha^4 + 2(D_{12} + 2D_{66})\alpha^2\beta^2 + D_{22}\beta^4 \\
s_{34} &= D^s_{11}\alpha^4 + 2(D^s_{12} + 2D^s_{66})\alpha^2\beta^2 + D^s_{22}\beta^4 \\
s_{44} &= H^s_{11}\alpha^4 + 2(H^s_{12} + 2H^s_{66})\alpha^2\beta^2 + H^s_{22}\beta^4 + A^s_{55}\alpha^4 + A^s_{44}\beta^4 \\
m_{11} &= I_0,\; m_{22} = I_0,\; m_{13} = -\alpha I_1,\; m_{14} = -\alpha J_1,\; m_{23} = -\beta I_1,\; m_{24} = -\alpha J_1 \\
m_{33} &= I_0 + (\alpha^2 + \beta^2)I_2,\; m_{34} = I_0 + (\alpha^2 + \beta^2)J_2,\; m_{44} = I_0 + (\alpha^2 + \beta^2)\hat{J}
\end{aligned}$$

The term F_{mn} in the load vector can be obtained as

$$F_{mn} = \frac{4}{ab}\int_0^b\int_0^a \hat{p}(x, y, z, t)\sin\alpha x \sin\beta y dxdy \tag{7.18}$$

The uniformly and sinusoidally distributed pressure – two kinds of loads – are considered as follows:

$$\begin{aligned}
&\text{For uniformly distributed load:} \\
&\hat{p}(x, y, z, t) = \hat{p}_0\hat{F}(t) \\
&\text{For sinusoidally distributed load:} \\
&\hat{p}(x, y, z, t) = \hat{p}_0 \sin\frac{\pi x}{a}\sin\frac{\pi y}{b}\hat{F}(t)
\end{aligned} \tag{7.19}$$

Substituting eq. (7.22) into eq. (7.21), the term F_{mn} is obtained as follows:

For uniformly distributed load:

$$F_{mn} = \frac{16\hat{p}_0}{mn\pi^2}\hat{F}(t) \tag{7.20}$$

For sinusoidally distributed load:

$$F_{mn} = \hat{p}_0\hat{F}(t)$$

We consider three of the function of times $\hat{F}(t)$ of dynamic loads as follows:

For sine loading

$$\hat{F}(t) = \begin{cases} \sin(\pi t/t_1) & 0 \le t \le t_1 \\ 0 & t \ge t_1 \end{cases}$$

For step loading

$$\hat{F}(t) = \begin{cases} 1 & 0 \le t \le t_1 \\ 0 & t \ge t_1 \end{cases}$$

For triangular loading

$$\hat{F}(t) = \begin{cases} t/t_1 & 0 \le t \le t_1 \\ 0 & t \ge t_1 \end{cases}$$

For convenience, the stiffness, mass, and forced matrices in eq. (7.17) are abbreviated as follows:

$$\boldsymbol{S} = \begin{bmatrix} s_{11} & s_{12} & s_{13} & s_{14} \\ s_{12} & s_{22} & s_{23} & s_{24} \\ s_{13} & s_{23} & s_{33} & s_{34} \\ s_{14} & s_{24} & s_{34} & s_{44} \end{bmatrix}, \quad \boldsymbol{M} = \begin{bmatrix} m_{11} & 0 & m_{13} & m_{14} \\ 0 & m_{22} & m_{23} & m_{24} \\ m_{13} & m_{23} & m_{33} & m_{34} \\ m_{14} & m_{24} & m_{43} & m_{44} \end{bmatrix}$$

$$\boldsymbol{F}^{*} = \{0 \quad 0 \quad F_{mn} \quad F_{mn}\}^{T}$$

Applying the state variable methods to solve eq. (7.17), they can be rewritten as

$$\dot{\boldsymbol{R}} = \boldsymbol{A}\boldsymbol{R} + \boldsymbol{b} \tag{7.21}$$

where

$$\boldsymbol{R} = \{U_{mn} \quad V_{mn} \quad W_{bmn} \quad W_{smn} \quad \dot{U}_{mn} \quad \dot{V}_{mn} \quad \dot{W}_{bmn} \quad \dot{W}_{smn}\}^{T}$$

$$\boldsymbol{b} = \{0 \quad 0 \quad 0 \quad 0 \quad \hat{b}_1 \quad \hat{b}_2 \quad \hat{b}_3 \quad \hat{b}_4\}^{T}$$

and $\hat{b}_1$ are the coefficients of vector force:

$$\left\{\hat{b}_i\right\} = \boldsymbol{M}^{-1}\boldsymbol{F}^{*}$$

The partitioned matrix $\boldsymbol{A}$ is defined as

$$\hat{\boldsymbol{A}} = \begin{bmatrix} \boldsymbol{0} & \boldsymbol{I} \\ -\boldsymbol{M}^{-1}\boldsymbol{S} & \boldsymbol{0} \end{bmatrix}$$

where $\boldsymbol{I}$ is the identity matrix.

Although the differential equation (7.21) is complex, we can find a solution by using the state variable method and the solution is obtained as follows:

$$\boldsymbol{R}(t) = e^{\hat{\boldsymbol{A}}(t-t_0)}\boldsymbol{R}(t_0) + \int_{t_0}^{t} e^{\hat{\boldsymbol{A}}(t-\tau)}\boldsymbol{b}(\tau)d\tau \tag{7.22}$$

where t_0 is the initial time and $\boldsymbol{R}(t_0)$ is the initial response while $e^{\hat{A}}$ is an exponential matrix; this matrix is created from eigenvalues λ_i and the eigenmatrix Λ of $\hat{\boldsymbol{A}}$ as follows:

$$e^{A(t-\tau)} = [\Lambda]\begin{bmatrix} e^{\lambda_1(t-\tau)} & & 0 \\ & \ddots & \\ 0 & & e^{\lambda_8(t-\tau)} \end{bmatrix}[\Lambda]^{-1} \tag{7.23}$$

With the initial condition at the time $t = 0$, it assumes that the system is stationary at the equilibrium position.

7.3 Results and discussions

In this section, the simple Al/Al_2O_3 plate made of aluminum and alumina is considered. The material properties used in the present study are chosen to be the same as [36]

$$\textit{Metal}: E_m = 70\text{ GPa},\ \rho_m = 2{,}707\text{ kg/m}^3$$

$$\textit{Ceramic}: E_c = 380\text{ GPa},\ \rho_c = 3{,}800\text{ kg/m}^3$$

The Poisson's ratio of the material is supposed to be equal to 0.3. The uniform distribution with the magnitude $\hat{P}_0$ is subjected on the plate Figure 7.2.

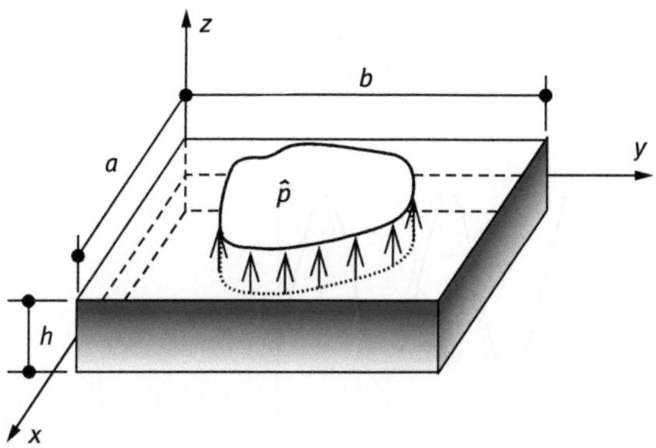

Figure 7.2: Outline of FG plate with the uniform load.

We consider the displacement at the center of the plate and normal stresses at the top and bottom of the center of the plate. For convenience, the normalized displacement and normalized stresses are given in the dimensionless form as

$$\tilde{w} = \frac{w}{h}$$

$$\tilde{\sigma}_x = \sigma_x \frac{\left(\frac{a}{2}, \frac{b}{2}, \pm\frac{h}{2}\right)}{\hat{p}_0}$$

$$\tilde{\sigma}_y = \sigma_y \frac{\left(\frac{a}{2}, \frac{b}{2}, \pm\frac{h}{2}\right)}{\hat{p}_0}$$

7.3.1 Comparison of the present approach with ABAQUS software

The square FG plate subjected to uniformly distribute dynamic loading with simply supported is performed as an example to verify the present analytical solution. The data for this example: dimensional plate $a = b = 0.3m$, $h = \frac{a}{10}$, the magnitude of the load $\hat{p}_0 = 10^6 N/m^2$, duration of load application time $t_1 = 0.0015s$, power-law index $k = 1$.

The ABAQUS commercial software is used to model FG plate subjected to dynamic load. The thickness direction of the FG plate is divided into fourteen layers. For modeling as composite layup, we use average properties within each layer. The time step for modeling dynamic load is $\Delta t = 1.5 \times 10^{-6}$ s. The results predicted by the present approach are compared with those obtained using ABAQUS software.

The normalized displacement and normalized stress at the central plate are given in Figures 3(a)–(c) and 4(a)–(c) due to triangular load and step load, respectively. It shows that the transient response of the FG plate using refined plate theory is in good

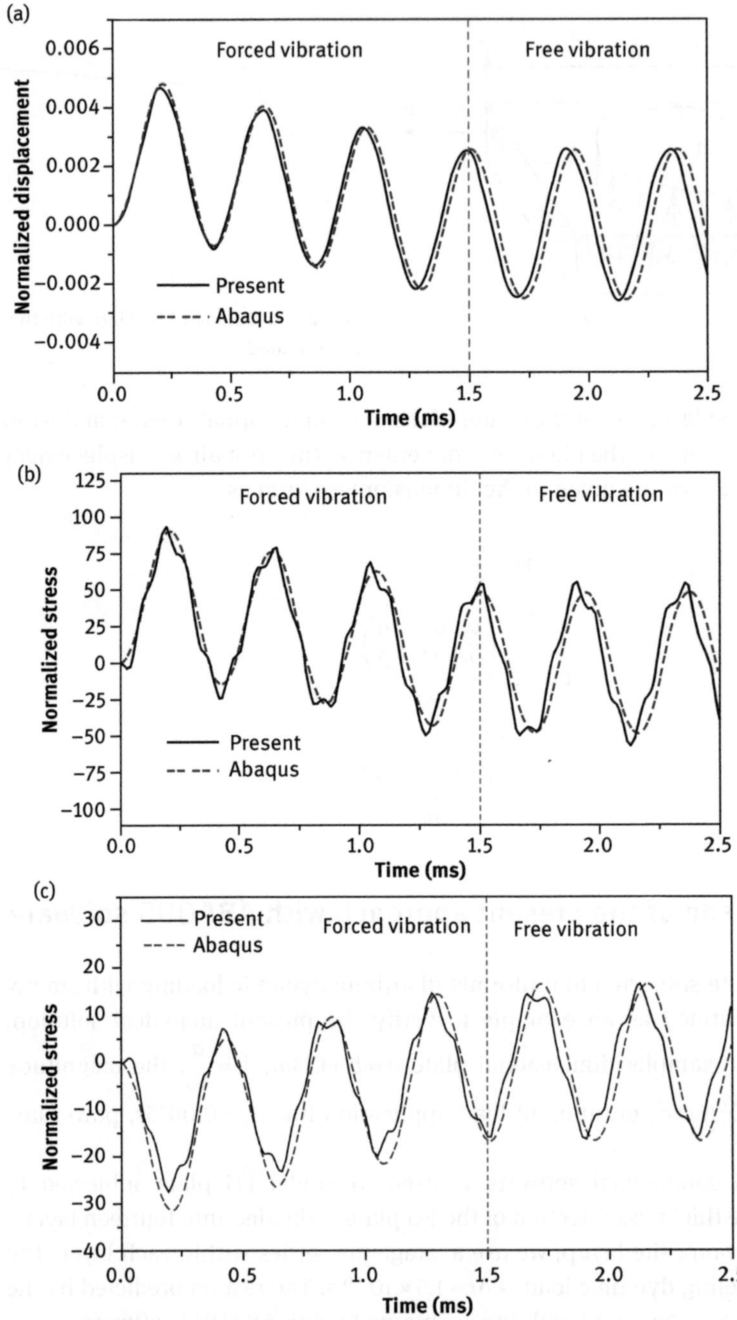

Figure 7.3: (a) Normalized displacement in case triangular load. (b) Normalized stress $\tilde{\sigma}_x$ ($z = h/2$) in case triangular load. (c) Normalized stress $\tilde{\sigma}_x$ ($z = -h/2$) in case triangular load.

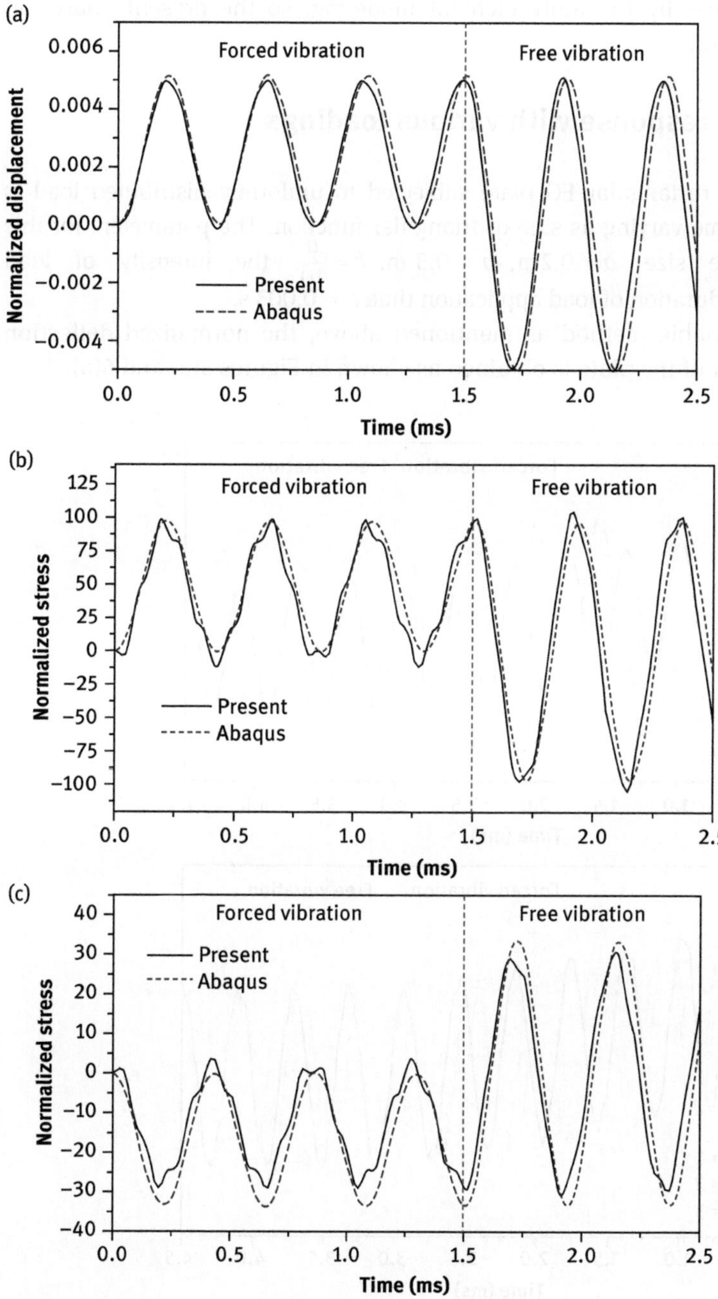

Figure 7.4: (a) Normalized displacement in case step load. (b) Normalized stress $\tilde{\sigma}_x$ $(z = h/2)$ in case step load. (c) Normalized stress $\tilde{\sigma}_x$ $(z = -h/2)$ in case step load.

agreement with those by the finite element modeling, so the present analytical solution is significant.

7.3.2 Transient response with various loadings

In this section, the rectangular FG plate subjected to uniformly distributed load is investigated with time varying as sine or triangular function. The parameters for this example are plate size $a = 0.2\,\text{m}$, $b = 0.3\ \text{m}$, $h = \dfrac{a}{10}$, the intensity of load $\hat{p}_0 = 10^6\ \text{N/m}^2$, and duration of load application time $t_1 = 0.003\ \text{s}$.

Using state variable method as mentioned above, the normalized deflection history at the center of the plate is obtained as shown in Figures 5(a) and 6(a).

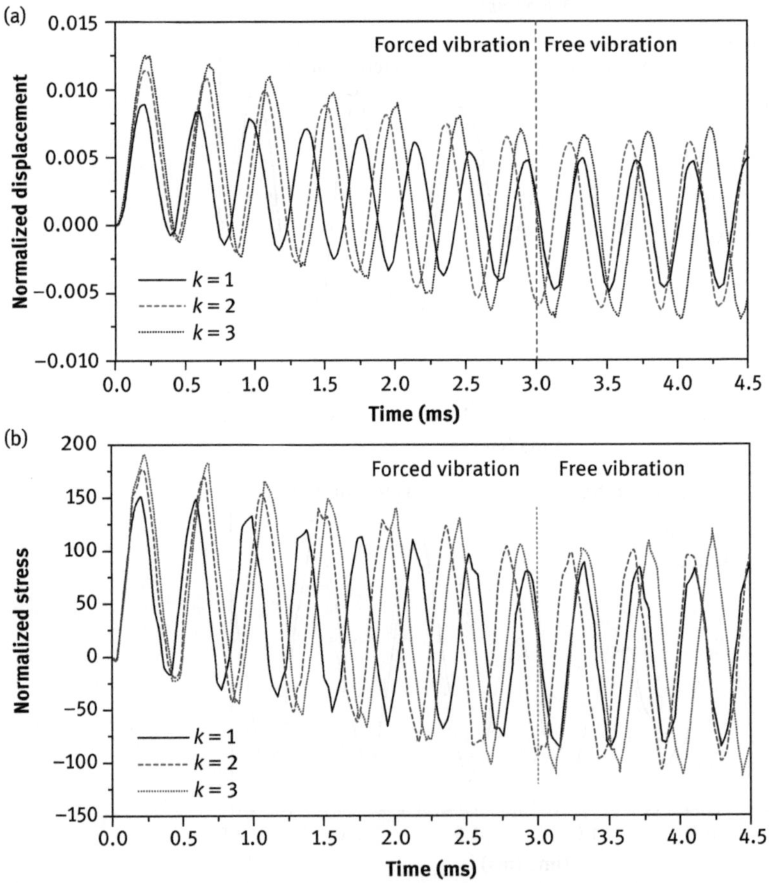

Figure 7.5: (a) Center deflection versus time due to triangular load. Normalized stress $\tilde{\sigma}_x$ $(z = h/2)$ in case sine triangular load. (c) Normalized stress $\tilde{\sigma}_y$ $(z = h/2)$ in case sine triangular load. (d) Normalized stress $\tilde{\sigma}_x$ $(z = -h/2)$ in case sine triangular load. (e) Normalized stress $\tilde{\sigma}_y$ $(z = -h/2)$ in case sine triangular load.

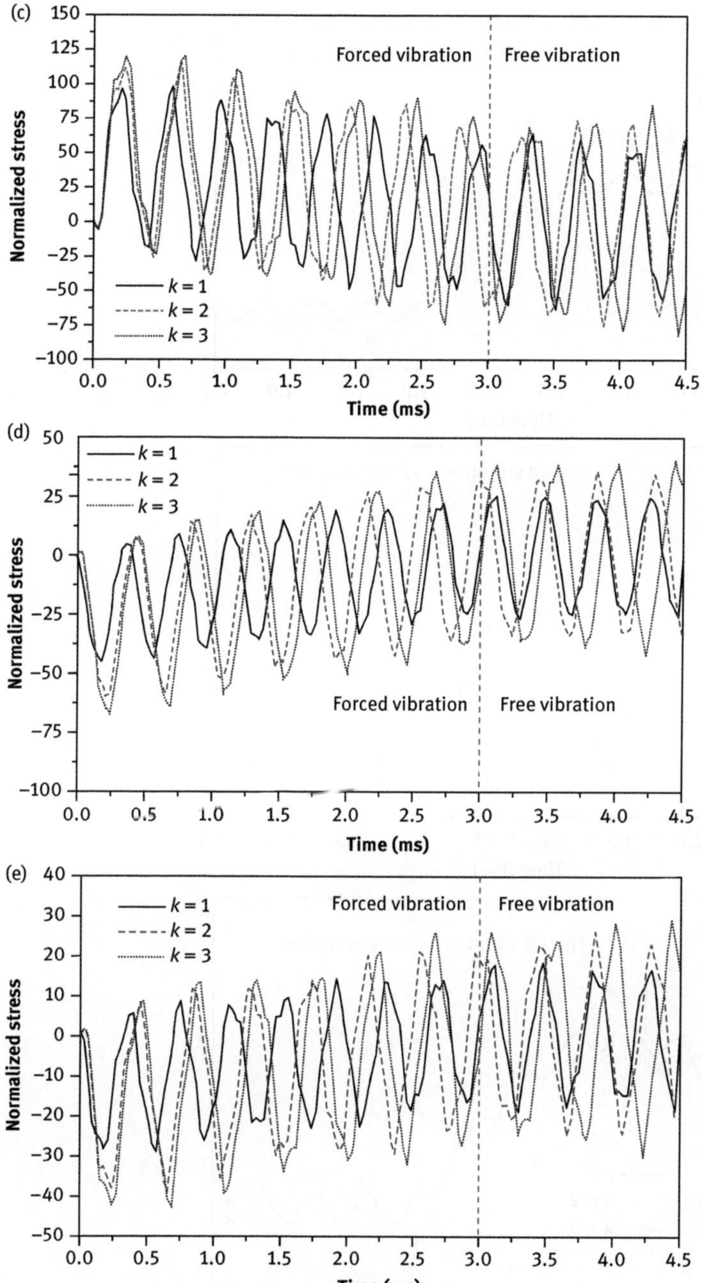

Figure 7.5: continued

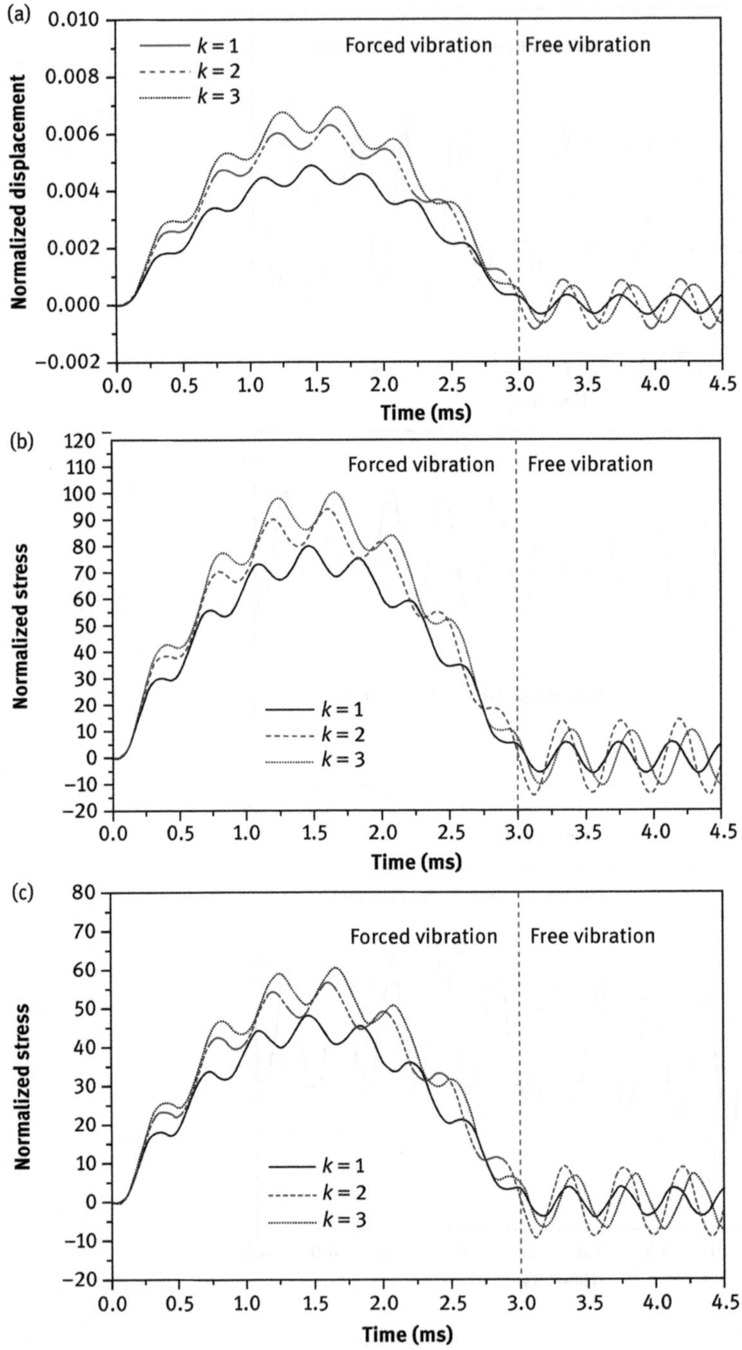

Figure 7.6: (a) Center deflection versus time due to sine load. (b) Normalized stress $\tilde{\sigma}_x$ $(z = h/2)$ due to sine load. (c) Normalized stress $\tilde{\sigma}_y$ $(z = h/2)$ due to sine load. (d) Normalized stress $\tilde{\sigma}_x$ $(z = -h/2)$ in case sine load. (e) Normalized stress $\tilde{\sigma}_x$ $(z = -h/2)$ in case sine load.

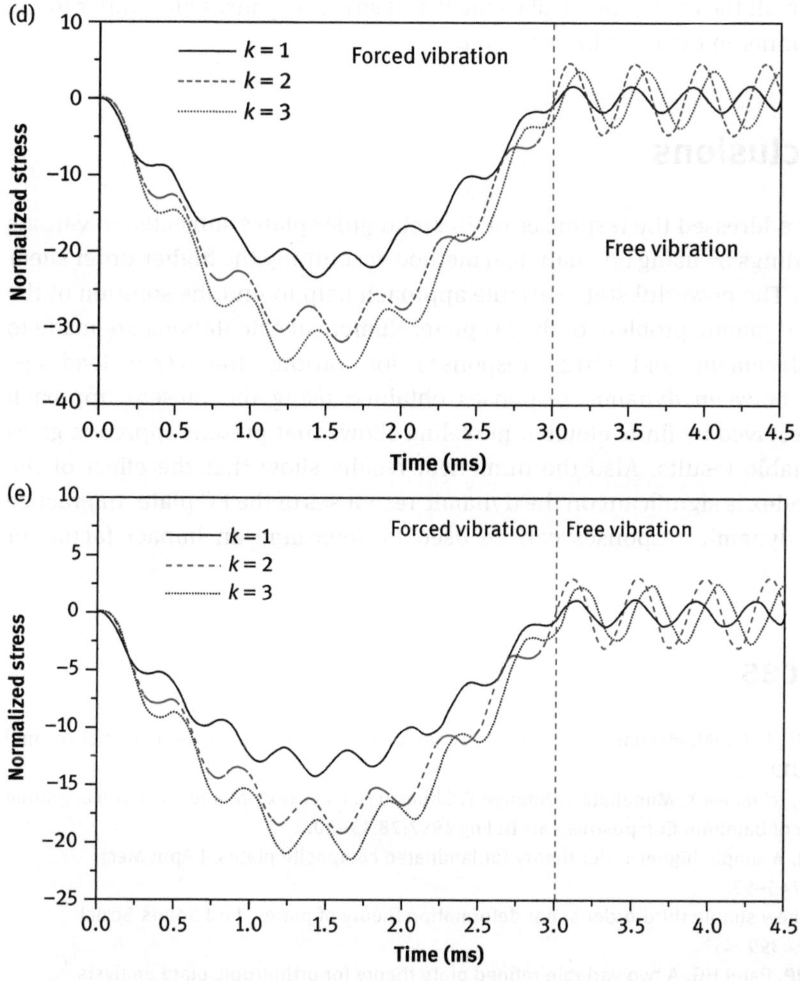

Figure 7.6: continued

Also, the responses of the normalized stresses at the central plate for triangular and step loads are illustrated in Figures 5(b)–(e) and 6(b)–(e). The numerical is presented with three cases of power-law index k, i.e., $k = 1$, $k = 2$, and $k = 3$. It showed in Figures 5 (a) and 6(a) that the deflection of central plate predicted for FG plate increases due to a decrease in flexural rigidity while power-law index k increases. In addition, from Figures 5(b)–(c) and 6(b)–(c), we can investigate the effect of the power-law index k on the normalized stress. The tendency of the normalized stresses is similar to the trend of the deflection, i.e., the normalized stress increases when power-law index k goes up. Moreover, in each load case, the normalized stress in x- or y-direction at a fiber has same shapes but the normalized stress in the x-direction is bigger than the

y-direction. In all the cases, the displacement and stress responses are similar to the harmonic manner in the free vibration area.

7.4 Conclusions

This chapter addressed the responses of FG rectangular plates subjected to various dynamic loadings by using an analytical method accounting the higher-order shear deformation. The powerful state variable approach help to find the solution of the complicated dynamic problem of the FG plate. Numerical calculations are made to obtain displacement and stress responses for various transverse loadings. Comparison between dynamic responses obtained using the present approach with those received by finite element modeling shows that present approach gives more reasonable results. Also the numerical results show that the effect of the power-law index is significant on the dynamic responses of the FG plate. In practice design, the dynamic responses can be used to determine an impact factor for design.

References

[1] Lakkad SC, Patel JM. Mechanical properties of bamboo, a natural composite. Fibre Sci Technol 1981;14:319–22.
[2] Amada S, Ichikawa Y, Munekata T, Nagase Y, Shimizu H. Fiber texture and mechanical graded structure of bamboo. Composites Part B: Eng 1997;28:13–20.
[3] Reddy JN. A simple higher-order theory for laminated composite plates. J Appl Mech 1984;51:745–52.
[4] Shi G. A new simple third-order shear deformation theory of plates. Int J Solids Struct 2007;44:4399–417.
[5] Shimpi RP, Patel HG. A two variable refined plate theory for orthotropic plate analysis. Int J Solids Struct 2006;43:6783–99.
[6] Lewiński T. On the twelfth-order theory of elastic plates. Mech Res Commun 1990;17: 375–82.
[7] Hosseini-Hashemi S, Rokni Damavandi Taher H, Akhavan H, Omidi M. Free vibration of functionally graded rectangular plates using first-order shear deformation plate theory. Appl Math Modell 2010;34:1276–91.
[8] Topal U, Uzman Ü. Free vibration analysis of laminated plates using first-order shear deformation theory. In: İnan E, Kırış A, editors. Vibration problems ICOVP 2005. Springer Netherlands, 2007:493–8.
[9] Naserian-Nik AM, Tahani M. Free vibration analysis of moderately thick rectangular laminated composite plates with arbitrary boundary conditions. Struct Eng Mech 2010;35:217–40.
[10] Bessaim A, Houari MS, Tounsi A, Mahmoud S, Bedia EA. A new higher-order shear and normal deformation theory for the static and free vibration analysis of sandwich plates with functionally graded isotropic face sheets. J Sandwich Struct Mater 2013;15:671–703.

[11] Hasani Baferani A, Saidi AR, Ehteshami H. Accurate solution for free vibration analysis of functionally graded thick rectangular plates resting on elastic foundation. Compos Struct 2011;93:1842–53.
[12] Muc A, Zuchara P. Sandwich plates-free vibrations and damping analysis. Mech Compos Mater 1998;34:203–11.
[13] Rajan LW, Kamal MB. Free vibration and stability analysis of piezolaminated plates using the finite element method. Smart Mater Struct 2013;22:125040.
[14] Binh BV, Thinh TI, Tu TM. Static and dynamic analyses of stiffened folded laminate composite plate. Vietnam Journal of Mechanics 2013;35:31–50.
[15] Taj MN, Chakrabarti A, Talha M. Free vibration analysis of four parameter functionally graded plates accounting transverse shear mode. Vietnam J Mech 2014;36:145–60.
[16] Bahmyari E, Banatehrani MM, Ahmadi M, Bahmyari M. Vibration analysis of thin plates resting on Pasternak foundations by element-free Galerkin method. Shock Vib 2013;20:309–26.
[17] Shojaee S, Izadpanah E, Valizadeh N, Kiendl J. Free vibration analysis of thin plates by using a NURBS-based isogeometric approach. Finite Elem Anal Des 2012;61:23–34.
[18] Shen H-S, Zheng JJ, Huang XL. Dynamic response of shear deformable laminated plates under thermomechanical loading and resting on elastic foundations. Compos Struct 2003;60:57–66.
[19] Huang XL, Zheng JJ. Nonlinear vibration and dynamic response of simply supported shear deformable laminated plates on elastic foundations. Eng Struct 2003;25:1107–19.
[20] Kazancı Z, Mecitoğlu Z. Nonlinear dynamic behavior of simply supported laminated composite plates subjected to blast load. J Sound Vib 2008;317:883–97.
[21] Rikards RB, Chate AK, Korjakin AV. Damping analysis of laminated composite plates by finite-element method. Mechanics of Compos Mater 1994;30:68–78.
[22] Ghafoori E, Asghari M. Dynamic analysis of laminated composite plates traversed by a moving mass based on a first-order theory. Compos Struct 2010;92:1865–76.
[23] Trung NT, Phuc PV, Anh TV, Chan NT. Dynamic analysis of Mindlin plates on viscoelastic foundations under a moving vehicle by cs-min3 based on c0-type higher-order shear deformation theory. Vietnam J Mech 2014;36:61–75.
[24] Susler S, Turkmen HS, Kazancı Z. The nonlinear dynamic behaviour of tapered laminated plates subjected to blast loading. Shock and Vibration 2012;19:1235–1255
[25] Hosseini-Hashemi S, Azimzadeh-Monfared M, Rokni Damavandi Taher H. A 3-D Ritz solution for free vibration of circular/annular functionally graded plates integrated with piezoelectric layers. Int J Eng Sci 2010;48:1971–84.
[26] Uymaz B, Aydogdu M, Filiz S. Vibration analyses of FG plates with in-plane material inhomogeneity by Ritz method. Compos Struct 2012;94:1398–405.
[27] Baferani AH, Saidi AR, Ehteshami H. On free vibration of functionally graded Mindlin plate and effect of in-plane displacements. J Mech 2013;29:373–84.
[28] Nguyen-Thoi T, Bui-Xuan T, Phung-Van P, Nguyen-Hoang S, Nguyen-Xuan H. An edge-based smoothed three-node Mindlin plate element (ES-MIN3) for static and free vibration analyses of plates. KSCE J Civil Eng 2014;18:1072–82.
[29] Tran LV, Ferreira AJ, Nguyen-Xuan H. Isogeometric analysis of functionally graded plates using higher-order shear deformation theory. Composites Part B: Eng 2013;51:368–83.
[30] Lei ZX, Liew KM, Yu JL. Free vibration analysis of functionally graded carbon nanotube-reinforced composite plates using the element-free kp-Ritz method in thermal environment. Compos Struct 2013;106:128–38.
[31] Yang J, Shen H-S. Dynamic response of initially stressed functionally graded rectangular thin plates. Compos Struct 2001;54:497–508.
[32] Yang J, Huang X-L. Nonlinear transient response of functionally graded plates with general imperfections in thermal environments. Comput Methods Appl Mech Eng 2007;196:2619–30.

[33] Wen PH, Sladek J, Sladek V. Three-dimensional analysis of functionally graded plates. Int J Numer Methods Eng 2011;87:923–42.
[34] Reddy JN. Analysis of functionally graded plates. Int J Numer Methods Eng 2000;47:663–84.
[35] Richard C. Dorf, Robert H. Bishop. Modern control Systems, 12 ed. Prentice Hall, New York. 2010.
[36] Benyoucef S, Mechab I, Tounsi A, Fekrar A, Ait Atmane H, Adda Bedia EA. Bending of thick functionally graded plates resting on Winkler–Pasternak elastic foundations. Mech Compos Mater 2010;46:425–34.

P. K. Karsh, T. Mukhopadhyay, and S. Dey

8 Fuzzy-based frequency response function analysis of functionally graded plates

Abstract: In the present chapter, fuzzy approach is utilized in the assessment of frequency response function (FRF) analysis of functionally graded plates (FGPs). FGP has significant applications in aerospace, marine, medical, and civil structures due to its unique properties such as high-temperature resistance, non-corrosiveness, and high strength and stiffness. The uncertainty quantification of FRF is portrayed. The fuzziness is considered due to its variability in material properties corresponding to the various α-cuts. The power law is implied for characterizing the material modelling. A parametric study is carried out to observe the effect of location of drive point and cross points on uncertain bounds of FRF with respect to crisp values.

Keywords: Fuzzy, frequency response function, functionally graded material, finite element, uncertain quantification

8.1 Introduction

Functionally graded materials (FGMs) are the hybrid of the composite, which has high-performance applications in various fields such as aerospace, marine, sports, biomedicals, and other structural applications. FGM is the inhomogeneous materials in which structures have specified properties in specified direction. The main feature of FGM is that desired material properties can be obtained by changing the volume fraction of constitute materials. This feature provides good corrosion and heat resistance [1]. FGM is used for manufacturing of heat shields, components of heat engine, fusion reactors, electric insulation, spacecraft structure, turbine wheels, leading edge of missile, cylinder liners of diesel engine, breaks of racing car, bullet proof vehicles, the solar cells, propeller shaft of submarines, and tennis rackets. Materials with single composition and uniform distribution cannot give such mechanical and thermal properties. The FGM is an advanced material in which two materials are mixed, and forms a nonhomogeneous mixture. There are many laws according to which material properties vary on the FGM such as power law index, sigmoid law, and exponential law. The main feature of FGM is that it has continuous and smooth variation of material properties throughout in depth. The mechanical and thermal properties of FGM is predetermined and it can be controlled by varying the composition of constitutes. In practical case, the variation of volume fraction in third direction (z-direction) is negligible. There is no sharp interface between boundaries of two different materials, and the resulting stress concentration is negligible [2]. In composite, two dissimilar materials with different

https://doi.org/10.1515/9783110545104-008

coefficient of thermal expansion are placed one above another, and when high temperature is applied, due to different expansion of both materials delamination of composite occurs. The main advantage of FGM over the composite is that it can resist more temperature and load as compared to composite, while due to delamination of layers composite fails under such conditions [3]. The basic idea of FGM came from nature, where already some natural things like teeth, wood, and bones are present [4]. FGM is made of two different materials, where the top layer is pure ceramic, and the bottom layer is pure metal. In between upper and lower surfaces of the plate, the volume fraction of materials changes continuously. The final properties of FGMs are totally different to the parent materials, which has properties of both materials simultaneously [5]. Each and every point of FGM has different mechanical, chemical, thermal, magnetic, and thermal properties. In metal–ceramic FGM, metal provides sufficient strength, while ceramic provides good thermal and corrosion resistance. There are many methods to manufacture FGM such as powder metallurgy, centrifugal method, chemical methods, and solid free-form fabrication method [6]. FGM can be classified based on the mechanical properties gradient, porosity gradients, microstructure gradient, and chemical gradient. In real-life situation, there are many types of uncertainty involved in the system such as manufacturing uncertainty, operational and environmental uncertainty, and geometric and material properties uncertainty. The manufacturing of FGMs always has some uncertainty in material properties in practical. The design and analysis of FGM are tougher than the conventional materials, because the variation of materials and geometrical properties of conventional materials from the nominal value are little or well known. For the safe and economical design of FGM structure, it is essential to consider an uncertain variation of material and geometrical properties [7–9]. Complex structures such as composite plates and shells have received wide attention from the research community for analysing the stochastic dynamic responses based on probabilistic approaches [10–13]. Effect of material and structural uncertainty on the dynamic stability and failure characteristics of composites are studied [14, 15]. The effect of uncertainty due to service conditions is accounted for composite structures [16, 17]. Effect of cut-out is analysed for the stochastic dynamic responses of composite structures [18]. Stochastic responses of sandwich structures and random honeycomb core are studied recently [19–23]. However, the influence of stochasticity in FGM structures has not received adequate attention yet.

8.2 Background

A plenty of research has been conducted to determine free vibration of FGM and composite. A review is conducted on finite element method (FEM) modelling for vibration analysis of composite [24]. The stochastic dynamic analysis of FGM beam is carried out by using random factor method [25], while FEM is applied for stochastic natural frequency analysis of FGM beam [26]. Free vibration of FGM is carried out by using Stochastic IsoGeometric Analysis method, and compared results with the Monto Carlo simulation (MCS). Young's modulus and mass density are considered as the random input parameters, and found that uncertainty of these material properties significantly affects the eigenvalue of the functionally graded plate (FGP) [27]. An analysis is conducted for non-linear free vibration for the FGP with elastic support carried out in the thermal environment. Material properties, volume fraction index, and support are considered as random input parameters [28]. A theoretical and numerical study is conducted for frequency response analysis of modified mechanical system [29]. A stability analysis of FGM pipe flowing the fluid is carried out by using the hybrid method to investigate the effect of fluid pressure, power law index, fluid velocity and damping on the stability of an FGM pipe. The results found that the stability of the FGM pipe is increased with increase in the power law exponents up to the value of 10. The divergence is introduced by the internal pressure of the fluid, while internal damping affects the dynamic behaviour of the system [30]. Frequency response function (FRF) analysis for updating the analytical model with frequency data selection and numerical balance is carried out [31], while FRF analysis of the sandwich plate with the viscoelastic layer is conducted by considering the effects of the physical and geometrical properties, material properties, and temperature [32]. Sparse adaptive polynomial chaos expansions are applied to the modified model for frequency analysis by using stochastic frequency transformation [33]. A statistical model is developed for FRF for a system by considering the uncertainty of the system [34], while Kriging model is applied for FRF analysis for model updating [35]. A bottom-up approach is applied for FRF analysis of laminate composite [8]. Dynamic analysis of FGM rectangular plate under motion is carried out [36]. Static and vibration of the sandwich plate is analysed by applying the mixed interpolation of tensorial components approach [37]. Nonlocal elasticity theory is applied to vibration analysis of piezoelectric FGM plates [38]. A numerical case study on FRF of damped system is conducted by using the fuzzy finite element (FE) procedure [39], in which uncertainty and variability of different parameters are taken into account and found that interval FEM is the more effective method to find the FRF as compared to the MCS. The polynomial chaos expansion is used to predict FRF for curve fitting along with the hypercube sampling for simple beam and aircraft wing [40], and compared the results with the MCS. A particle swarm method with the fuzzy controller with integrated self-tuning is applied to free vibration control of composite beams [41], while a fuzzy uncertainty propagation approach for composites is proposed based on Gram–Schmidt polynomial chaos expansion [42]. The meshless local Petrov–Galerkin method is applied for three-dimensional vibration analysis of isotropic FGM plates. The three-dimensional linear elasticity theory is used, where it has high accuracy as compared to two-dimensional theory. The Young's modulus of the FGM plate is varied according to the Mori–Tanaka, and Poisson's ratio is kept constant [43]. Frequency responses are optimized for probability distribution function inspired structure [44]. The Papkovich–Neuber potentials are applied for frequency analysis of multilayer coating, which is mostly used for surface modification by analytical method, when subjected to normal and shear loading [45]. An FGM plate subjected to moving mass is analysed for free vibration by using the hybrid Rayleigh–Ritz method. The advantage of this method is that for different boundary condition, it is not required to determine the satisfied admissible function and also this method can be applied for versatile boundary condition. The effect of different boundary condition and material properties on maximum deflection are determined [46]. The resonance modes method is applied to determine the time and frequency response of aluminium plate, and found that the presented method has high efficiency and precision [47]. Energy method, Hamilton's principle, and Broyden method are used to determine the forced vibration of non-uniform FGM plate subjected to harmonic excitation for different boundary conditions, and taper parameter. The results concluded that the rigid boundary conditions have less response amplitude, and also response amplitude decreases with increase in taper parameters [48]. A functionally graded reinforced composite is statically and dynamically analysed by using the analytical method. First-order share deformation theory and Hamilton principle are used [49]. The FRF is an important structural parameter which is expressed as the ratio of response output of a structure due to an applied force. Such a modal response can be measured in the form of displacement, velocity, and acceleration. Uncertainty in material properties are modelled by using non-probabilistic and probabilistic approach. There are limited amount of sample data available which is also known as testing data (crisp inputs), so it is better to use non-probabilistic approach in place of probabilistic approach. Due to limited number of experimental data and impreciseness in input parameters, the concept of fuzzy is introduced in the present chapter. Hence, a fuzzy-based approach is used to quantify the uncertainty and their effects on the FRFs of FGM cantilever plates. Figure 8.1 shows the geometric view of FGM cantilever plate.

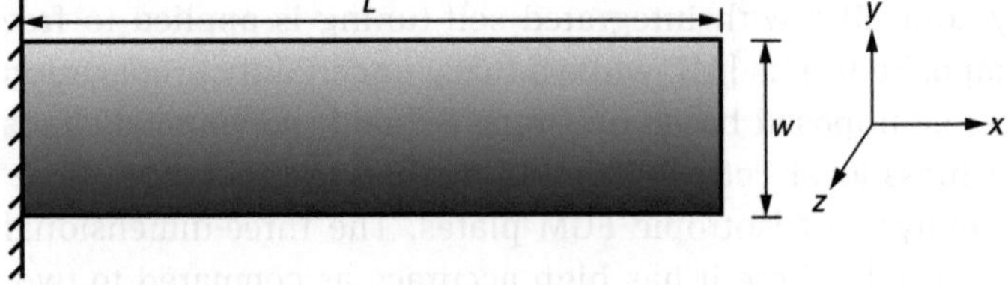

Figure 8.1: Geometry of FGM cantilever plate.

8.3 Governing equations

If “A” represents the function of material properties and V_i represents the volume fraction of individual materials “i”, then “A” can be given as

$$A = \sum_{i=1}^{n} A_i V_i \tag{8.1}$$

where A_i denotes the material property of the constituent material “i”.

In FGP, the material properties are considered to vary in the entire domain across the thickness [50]. In the present chapter, power law is implied to obtain the deterministic material properties of FGPs and expressed as

$$E(w) = E_m + (E_c - E_m)\left[\frac{y}{w} + \frac{1}{2}\right]^p \tag{8.2}$$

$$\mu(w) = \mu_m + (\mu_c - \mu_m)\left[\frac{y}{w} + \frac{1}{2}\right]^p \tag{8.3}$$

$$\rho(w) = \rho_m + (\rho_c - \rho_m)\left[\frac{y}{w} + \frac{1}{2}\right]^p \tag{8.4}$$

where E_m and E_c are Young’s modulus, μ_m and μ_c are Poisson’s ratio, and ρ_m and ρ_c are the volumetric mass density for metal and ceramic, respectively, while “w” denotes the thickness of plate, and $y = (w/2)$. Power law index is donated by “p”. The material properties of components of FGM are considered as $E_m = 70$ GPa, $E_c = 151$ GPa, $\mu_m = 0.25$, $\mu_c = 0.3$, $\rho_m = 2{,}707$ (kg/m^3) and $\rho_c = 3{,}000$ (kg/m^3) at 300 K [51].

After solving the eigenvalue problem, natural frequencies (ω_f) and mode shapes (S_f) can be determined [52]:

$$[K(\hat{\omega}_\alpha)]S_f = \omega_f^2 [M(\hat{\omega}_\alpha)]S_f \tag{8.5}$$

where $f = 1, 2, 3, \ldots, n$, $[K(\hat{\omega})]$ and $[M(\hat{\omega})]$ are the fuzzified stiffness and mass matrices. The symbol $(\hat{\omega})$ indicates the degree of fuzziness. The orthogonality relationship is satisfied by eigenvalue and eigenvector, which can be expressed as

$$S_i^T[M(\hat{\omega}_\alpha)]S_f = \lambda_{ik} \text{ and } S_i^T[K(\hat{\omega}_\alpha)]S_f = \omega_f^2 \lambda_{ik} \tag{8.6}$$

where $i, f = 1, 2, 3, \ldots, n$.

The Kronecker delta functions $\lambda_{ik} = 0$ for $i \neq k$ and $\lambda_{ik} = 1$ for $i = k$ wherein the relationship of orthogonality is expressed as

$$B^T[M(\hat{\omega}_\alpha)]B = I \text{ and } B^T[K(\hat{\omega}_\alpha)]B\Phi^2 \tag{8.7}$$

where B denotes the undamped nodal matrix and I represents an identity matrix with size$(n \times n)$. For a linear damped system, the equation of motion is expressed as

$$[M(\hat{\omega}_\alpha)]\{\ddot{\delta}_e\} + [C]\{\dot{\delta}_e\} + [K_e(\hat{\omega}_\alpha)]\{\delta_e\} = f(t) \tag{8.8}$$

where $[C]$ represents a damping coefficient matrix. Coordinate transformation is given as

$$\delta(\hat{\omega}_\alpha)(t) = By(t) \tag{8.9}$$

In the model coordinates, an equation of motion for the damped system is given by

$$\ddot{y}(t) + B^T CB\dot{y}(t) + \Phi^2 y(t) = \tilde{f}(t) \tag{8.10}$$

The damping matrix is expressed by eq. (8.11), which is a combination of mass matrix and stiffness matrix as follows:

$$C(\hat{\omega}_\alpha) = \psi M(\hat{\omega}_\alpha) f(M^{-1}(\hat{\omega}_\alpha) K(\hat{\omega}_\alpha)) \tag{8.11}$$

where ψ represents the constant damping factor, which has the value equal to 0.005.

The transfer function matrix for a given system is expressed as [53]

$$R(i\omega)(\hat{\omega}_\alpha) = B\left[-\omega^2 I + 2i\omega\zeta\Phi + \Phi^2\right]^{-1} B^T = \sum_{j=1}^{n} \frac{B_j B_j^T}{-\omega^2 + 2i\omega\zeta_j\omega_j + \omega_j^2} \tag{8.12}$$

where ζ denotes the local coordinate of the system. The dynamic behaviour of the system in the frequency domain considering zero initial conditions is expressed by

$$\tilde{\delta}(i\omega)(\hat{\omega}_\alpha) = R(i\omega)\tilde{f}(i\omega) = \sum_{j=1}^{n} \frac{B_j^T \tilde{f}(i\omega)}{-\omega^2 + 2i\omega\zeta_j\omega_j + \omega_j^2} B_j \tag{8.13}$$

so that the dynamic behaviour of the damped system can be represented by the combination of the undamped mode shapes.

8.4 Fuzzy-based approach

In this chapter, the pioneering concept of fuzzy [54] is utilized in conjunction with material property gradation and subsequently FRF bounds are mapped. The figure 8.2

shows the steps for fuzzy based frequency response analysis of FGM plates. The explanation of mathematical description of fuzzy variables is carried out by different levels of abstraction. Interval variables are the easiest way to explain the fuzzy variables. The difference between members and non-members is distinguished by a classical set; a degree of membership is obtained by the fuzzy set known as the membership function. The fuzzy number denotes a fuzzy set which is considered as convex, and at one element the membership function is a continuum at the functional value equal to 1. The fuzzy set is considered as an extension of a conventional (crisp) set, which discriminates between elements that belong to the set and those which do not. A membership function $[\mu_{y(i)}]$ is used to define the set of transformation phases between the non-members and members. Experimental data or expert knowledge is used to derive the shape of membership function. There are two popular types, namely triangular and Gaussian shapes. The fuzzy number $[\tilde{Y}_i(\hat{\omega}_\alpha)]$ considered as triangular membership function can be expressed as

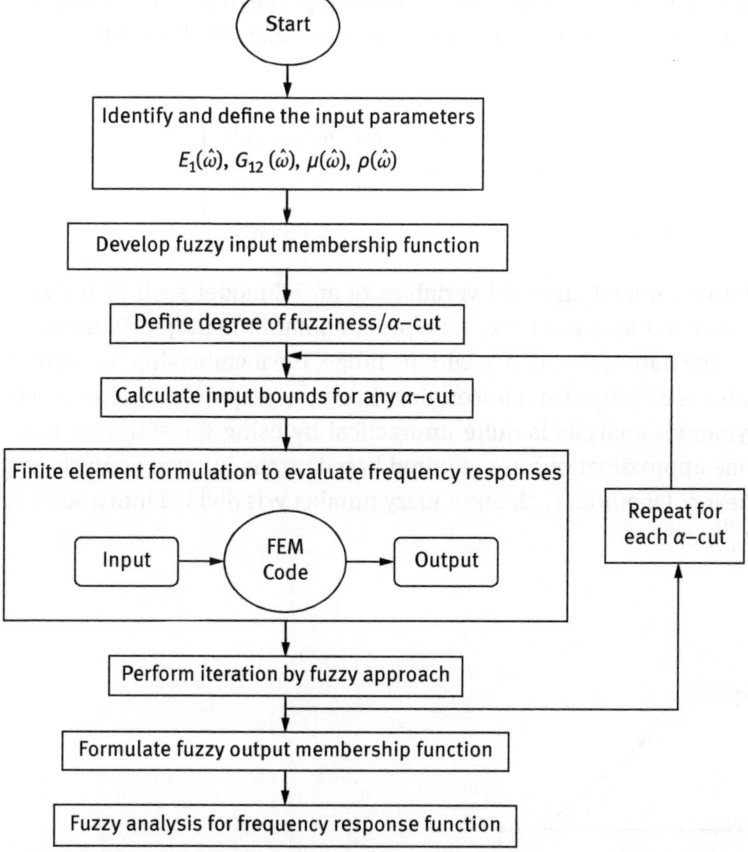

Figure 8.2: Flowchart for fuzzy-based frequency responses.

$$\tilde{Y}_i(\hat{\omega}_\alpha) = [y_i^U, y_i^M, y_i^L] \tag{8.14}$$

where y_i^M is the mean value while y_i^U is the upper bound and y_i^L is the lower bounds. $\hat{\omega}_\alpha$ shows the fuzziness corresponding to α-cut, where α is the degree of fuzziness or membership grade ranging from 0 to 1.

The approximation of Gaussian distribution can be carried out by using a triangle [56], wherein the area covered by the normalized Gaussian distribution function and triangular membership function is equated as illustrated in Figure 8.3. The linear approximated triangular fuzzy membership function is expressed as

$$\mu_{y(i)} = \max\left[0, 1 - \frac{\left|C_i^{(j)} - \hat{C}_i\right|}{\zeta}\right] \tag{8.15}$$

where $\zeta = \sqrt{2\pi}\sigma_X$, $\hat{C}_i$ denotes the mean value and σ_X is the standard deviation of the approximated Gaussian distribution. The membership function with triangular shape is used in the present chapter. The triangular membership function $[\mu_{y(i)}]$ is given by

$$\mu_{y(i)} = \begin{Bmatrix} 1 - (y_i^M - y_i)/(y_i^M - y_i^L), & \text{for } y_i^L \le y_i \le y_i^M \\ 1 - (y_i - y_i^M)/(y_i^U - y_i^M), & \text{for } y_i^M \le y_i \le y_i^U \\ 0 & \text{Otherwise} \end{Bmatrix} \tag{8.16}$$

In the present chapter, uncertain input variables of an FE model such as material properties are considered as input fuzzy numbers to obtain the output fuzzy numbers. To determine the natural frequency with its range, the membership function of output fuzzy number is employed and uncertainty in output responses is determined. The structural dynamics analysis is quite unpractical by using the standard fuzzy arithmetic, so some approximations are required based on the interval analysis and α-cut. By using the α-cut method, each input fuzzy number y_i is divided into a set Y_i of $(h + 1)$ intervals $C_i^{(j)}$,

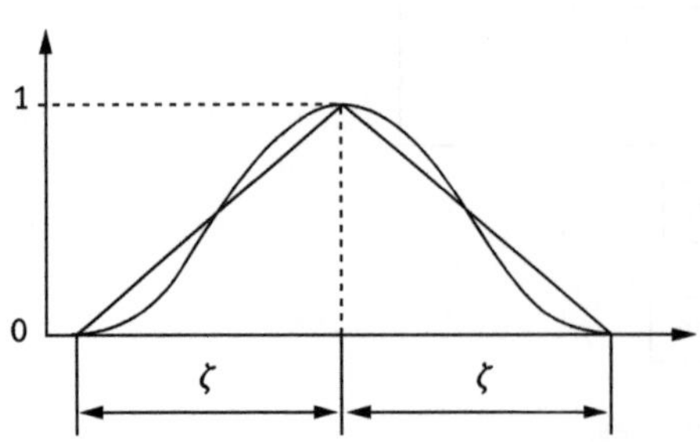

Figure 8.3: Approximated Gaussian distribution by using the triangular fuzzy [55].

$$Y_i(\hat{\omega}_\alpha) = [C_i^{(0)},\ C_i^{(1)}, C_i^{(2)} \ldots C_i^{(j)} \ \ldots \ C_i^{(h)}] \tag{8.17}$$

where h represents the number of levels of α-cut. For jth level of the ith fuzzy number, the interval is expressed as

$$C_i^{(j)} = [y_i^{(j,L)}, y_i^{(j,U)}] \tag{8.18}$$

where $y_i^{(j,L)}$ and $y_i^{(j,U)}$ are intervals at the jth level for lower bound and upper bound, respectively, and $y_i^{(h,L)} = y_i^{(h,U)} = y_i^M$ at $j = h$.

The fuzzy analysis uses the applied numerical procedure of interval analysis employed for FGPs considering "n" number of layers at a number of α-levels as illustrated in Figure 8.4. Figure 8.4 shows the procedure for a function of two triangular fuzzy parameters with four α-levels. For the particular level of the membership function, the range of response is determined within the same α-cut level on the input domain. In the interval analysis, the intervals of input fuzzy numbers are used to

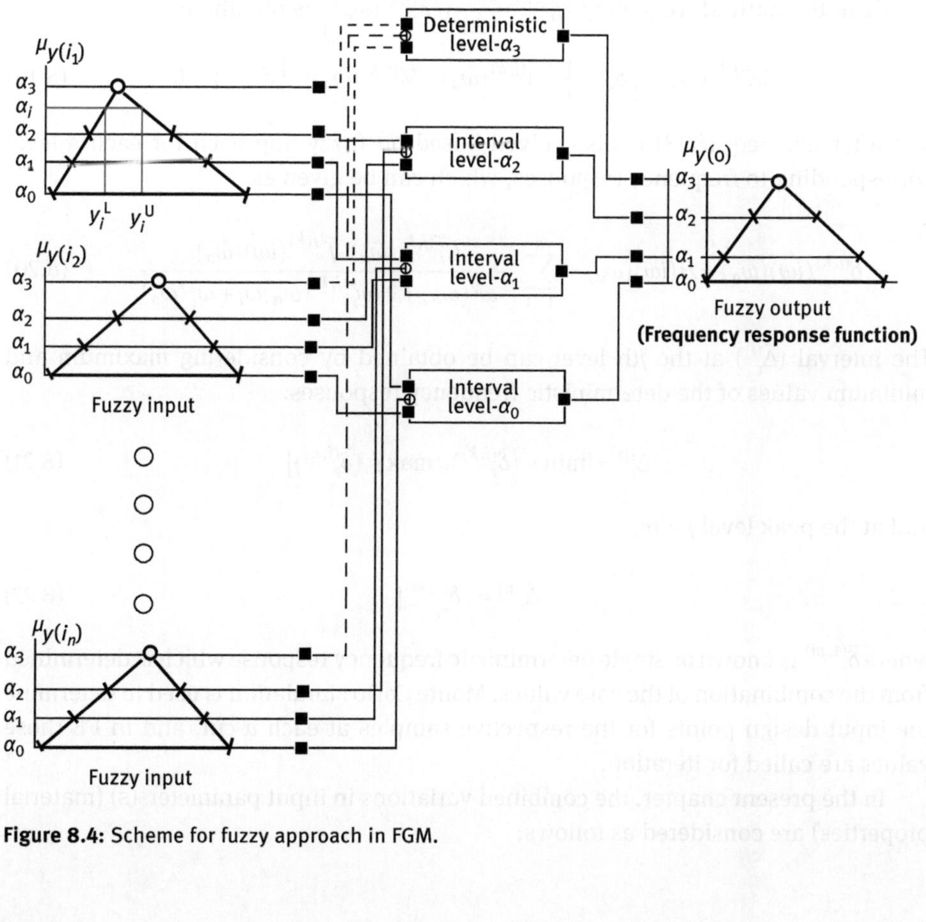

Figure 8.4: Scheme for fuzzy approach in FGM.

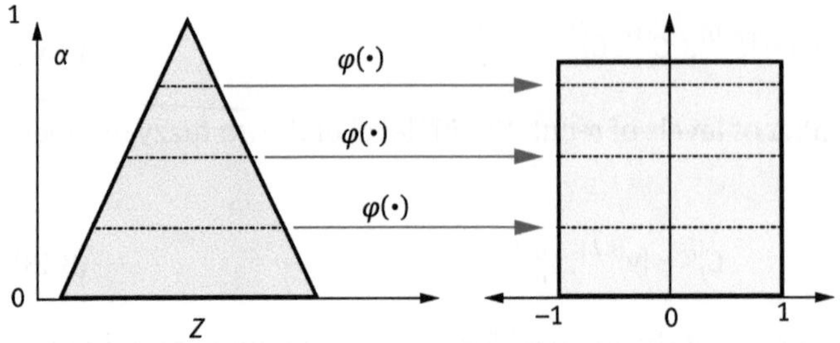

Figure 8.5: Transformation $\phi(\bullet)$ of a fuzzy variable z to $\xi \in [-1, 1]$ for different α-cuts [57].

determine the interval of output fuzzy number at each α-level, and based on the interval levels, the membership function of output fuzzy number is created by recomposition of intervals. Figure 8.5 shows the variation of fuzzy input variables of FGM.

The mass matrix $[M^{(j,k)}(\hat{\omega}_\alpha)]$ and stiffness matrix $[K^{(j,k)}(\hat{\omega}_\alpha)]$ are determined with the parameter values at the kth design point. By solving the eigenvalue problem, the natural frequency $\left[f_r^{(j,k)} = \frac{1}{2\pi}\sqrt{\lambda_r^{(j,k)}(\hat{\omega}_\alpha)}\right]$ is obtained:

$$[K^{(j,k)}(\hat{\omega}_\alpha)]\left\{\delta_r^{(j,k)}\right\} - \lambda_r^{(j,k)}(\hat{\omega}_\alpha)\,[M^{(j,k)}(\hat{\omega}_\alpha)]\left\{\delta_r^{(j,k)}\right\} = 0 \tag{8.19}$$

and after that, eq. (8.13) is also solved based on fuzzy approach for each α-level corresponding to frequency responses, which can be given as

$$\bar{\delta}^{(j,k)}(i\omega)(\hat{\omega}_\alpha) = H(i\omega)(\hat{\omega}_\alpha) = \sum_{i=1}^{n} \frac{C_i^{T(j,k)}(\hat{\omega}_\alpha)\,\bar{f}^{(j,k)}(i\omega)(\hat{\omega}_\alpha)}{-\omega^2(\hat{\omega}_\alpha) + 2i\omega\zeta_i^{(j,k)}(\hat{\omega}_\alpha)\omega_i + \omega_i^2(\hat{\omega}_\alpha)} \tag{8.20}$$

The interval $(\bar{\Delta}_r^{(j)})$ at the jth level can be obtained by considering maximum and minimum values of the deterministic frequency responses:

$$\bar{\Delta}_r^{(j)} = [\min_k\,(\bar{\delta}_r^{(j,k)}),\ \max_k\,(\bar{\delta}_r^{(j,k)})] \tag{8.21}$$

and at the peak level $j = m$,

$$\bar{\Delta}_r^{(m)} = [\,\bar{\delta}_r^{(1,m)}] \tag{8.22}$$

where $\bar{\delta}_r^{(1,m)}$ is known as single deterministic frequency response which is determined from the combination of the core values. Monte Carlo simulation is used to determine the input design points for the respective samples at each α-cut, and in FE those values are called for iteration.

In the present chapter, the combined variations in input parameters(s) (material properties) are considered as follows:

Combined variation of material properties $g\{E_1(\bar{\omega}), G_{12}(\bar{\omega}), \mu(\bar{\omega}), \rho(\bar{\omega})\} = \{\Phi_1(E_{1(1)}, E_{1(2)}, \ldots, E_{1(i)}), \Phi_2(G_{12(1)}, G_{12(2)}, \ldots, G_{12(i)}), \Phi_3(\mu_{(1)}, \mu_{(2)}, \ldots, \mu_{(i)}), \Phi_4(\rho_{(1)}, \rho_{(2)}, \ldots, \rho_{(i)})\}$

where E_1 = longitudinal Young's modulus, G_{12} = longitudinal modulus of rigidity, μ = Poisson's ratio, and ρ = volumetric mass density.

8.5 Results and discussion

An isoparametric quadratic element with eight nodes and each node having five degrees of freedom is considered in the FE formulation wherein mesh size adopted as (6 × 6). The present FE model is validated with the past literature ([58] and [59]) as furnished in Table 8.1. The concept of fuzzy is applied in conjunction with the FE model to quantify the uncertainties of FRF. The deterministic FRF for a system shows the mathematical relationship between the input and the output, wherein a force gauge hammer is employed for excitation, and an accelerometer measures the response. The point at which load is applied is called drive point while points at which response are to be calculated are known as cross points. Figure 8.6 illustrates the variation of material properties across the depth of plate based on fuzzy membership grade. The different frequencies of model are represented by the peaks of the FRF. Figure 8.7 shows the location of drive point and cross points. Figure 8.8 shows the effect of percentage of variation on fuzzy bounds and crisp values for amplitude (dB) with respect to frequency (rad/s) of cross points (C_1, C_2, C_3) and drive point considering the combined variation of material properties of FGM cantilever plate. It is concluded from Figure 8.8 as the percentage of variation (p) increases, fuzzy bound increased irrespective of the point. It is also noted that the values of fuzzy bound and crisp value obtained at cross points C_2 and C_3 are the same. This is due to the symmetricity of these points. The value of fuzzy bound at drive point is more as compared to the cross points. It is clear from Figure 8.8 that the percentage of variation (p) have no effect on the crisp valve irrespective of points. Figure 8.9 illustrates the compound effect of variation of input parameters

Table 8.1: Dimensionless first natural frequency $\left[\omega = \omega\sqrt{\frac{\rho_m}{E_m}}\right]$ of FGP with considering power law exponent (p) and thickness–length ratio (w/l).

p	*w/l*	Hasani et al. 2011	[58]	Present FEM
0	0.1	0.1134	0.1134	0.1139
	0.2	0.4154	0.4152	0.4159
1	0.1	0.0891	0.0869	0.0883
	0.2	0.3299	0.3205	0.3261
2	0.1	0.0819	0.0788	0.0797
	0.2	0.3016	0.2897	0.3004

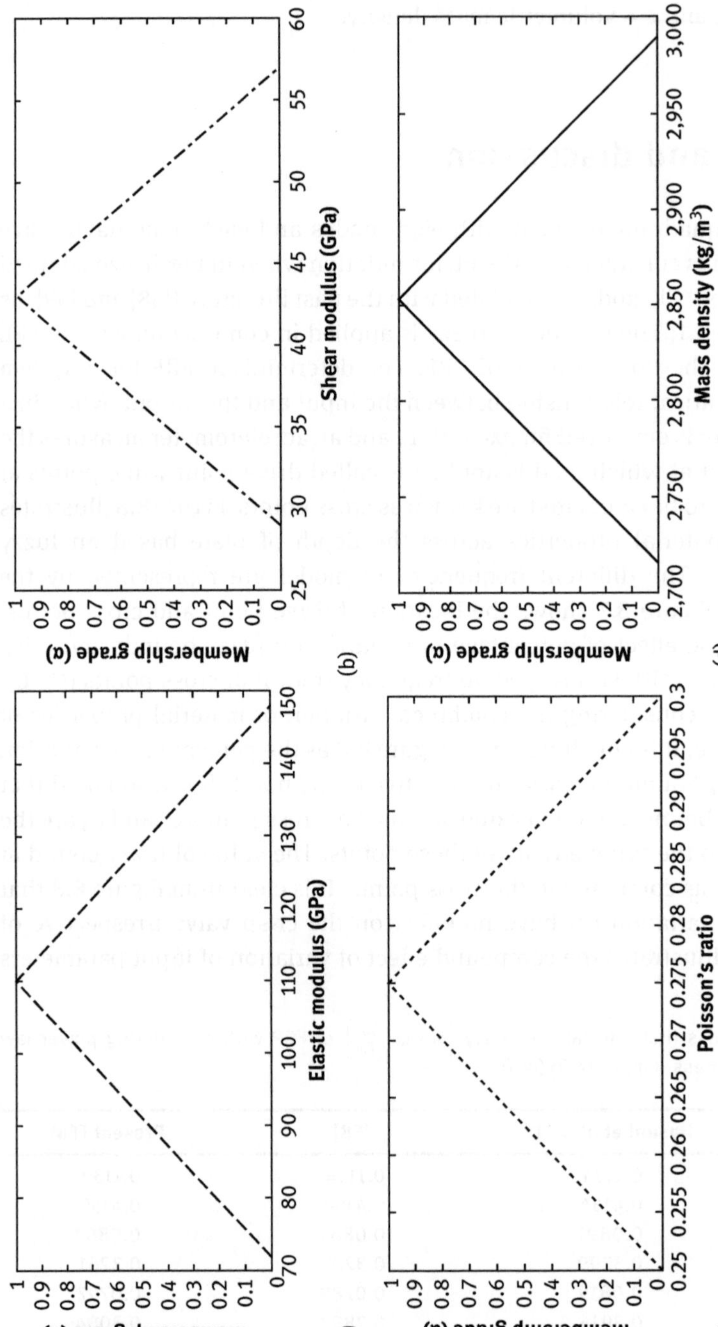

Figure 8.6: Variation of fuzzy input parameters for FGM cantilever plates.

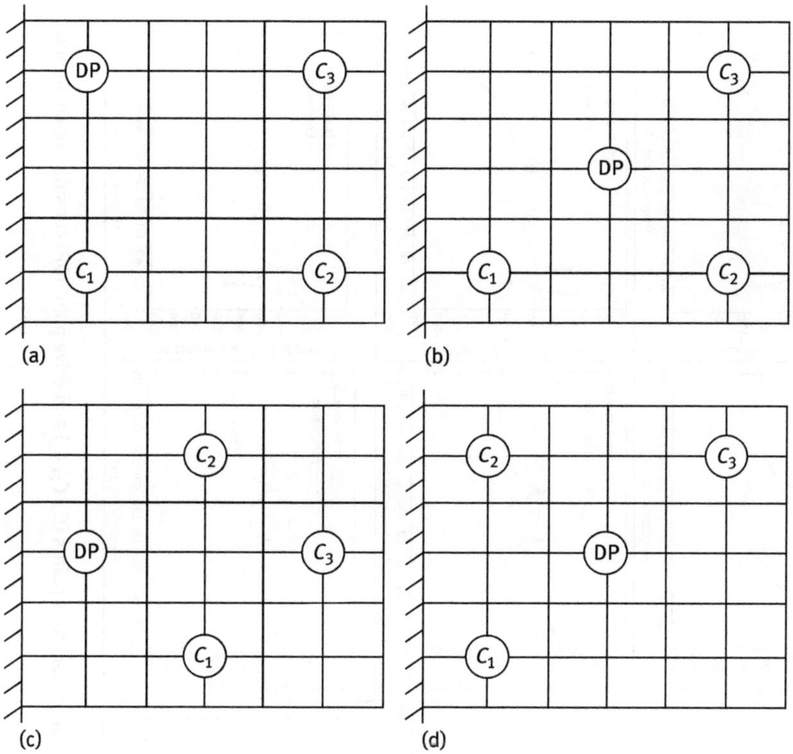

Figure 8.7: Location of drive point and cross point.

on the frequency response at different α-cuts. The minimum fuzzy bounds are found at $\alpha = 0.2$. As α value increases, fuzzy bounds first increases, and then decreases. At $\alpha = 0.2$, Figure 8.9 shows only crisp value without any fuzzy bounds. Three case studies are conducted to determine the effect of location of drive point and cross points on the fuzzy bound and crisp value. From Figure 8.10 it is concluded that when changing the location of drive point or cross points, the fuzzy bound and crisp value also changes.

In this chapter, the first attempt is carried out to find the FRF of functionally graded cantilever plates by using fuzzy. The present fuzzy-based approach can be extended for more complex structure for such analysis. However, the main challenge still remains in the computational intensiveness of the analysis. As direct closed-form solutions [60–63] are not available for the dynamic responses of complex structures, a surrogate-based approach can be adopted to achieve computational efficiency [64–70]. Further, the future work is needed in the areas of environmental effect and service life conditions of FGM structures considering non-deterministic approaches.

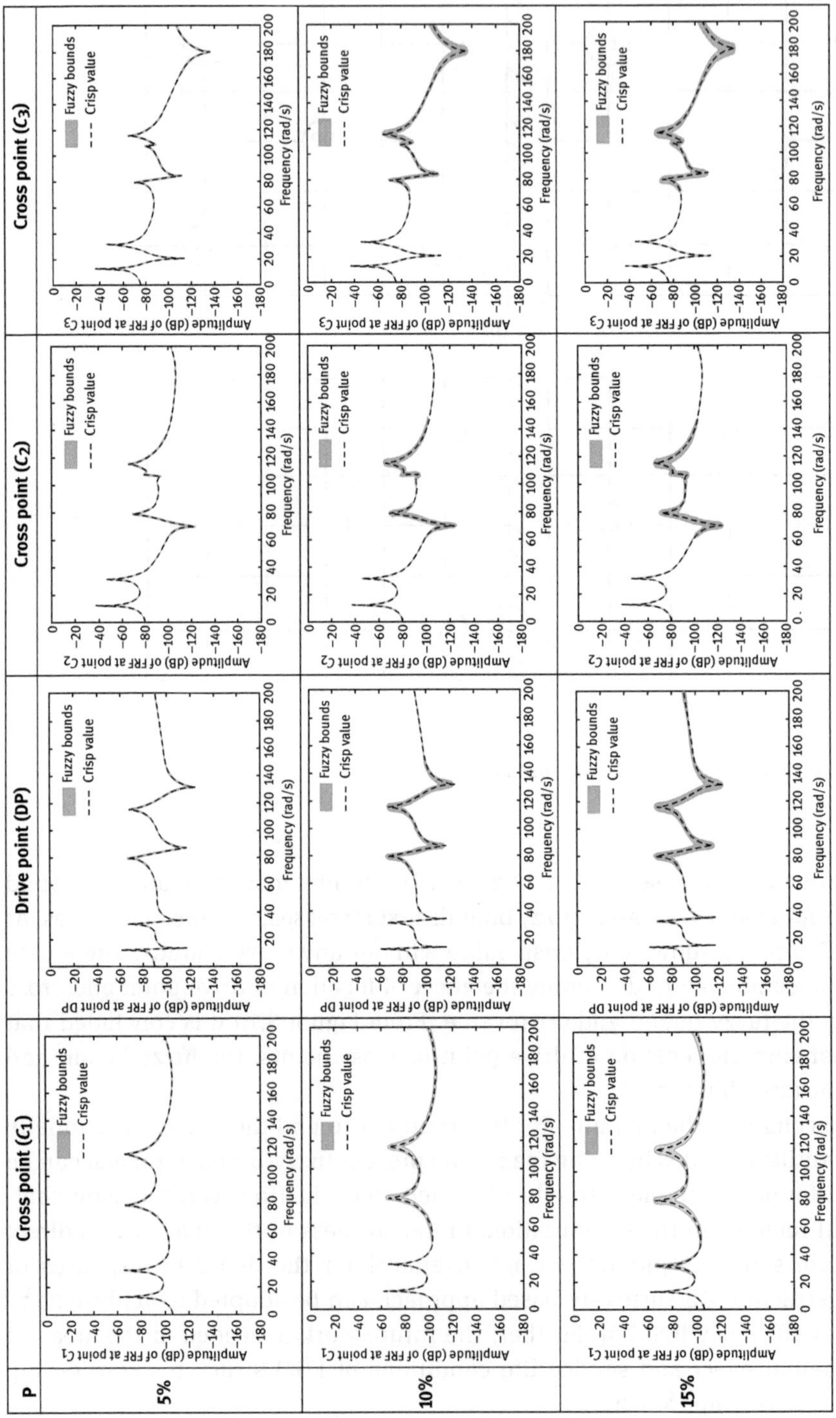

Figure 8.8: Effect of percentage of variation (p) on fuzzy bounds and crisp values for cross points (C_1, C_2, C_3) and drive point (DP) considering combined variation of material properties of FGM cantilever plate for Figure 8.6(a).

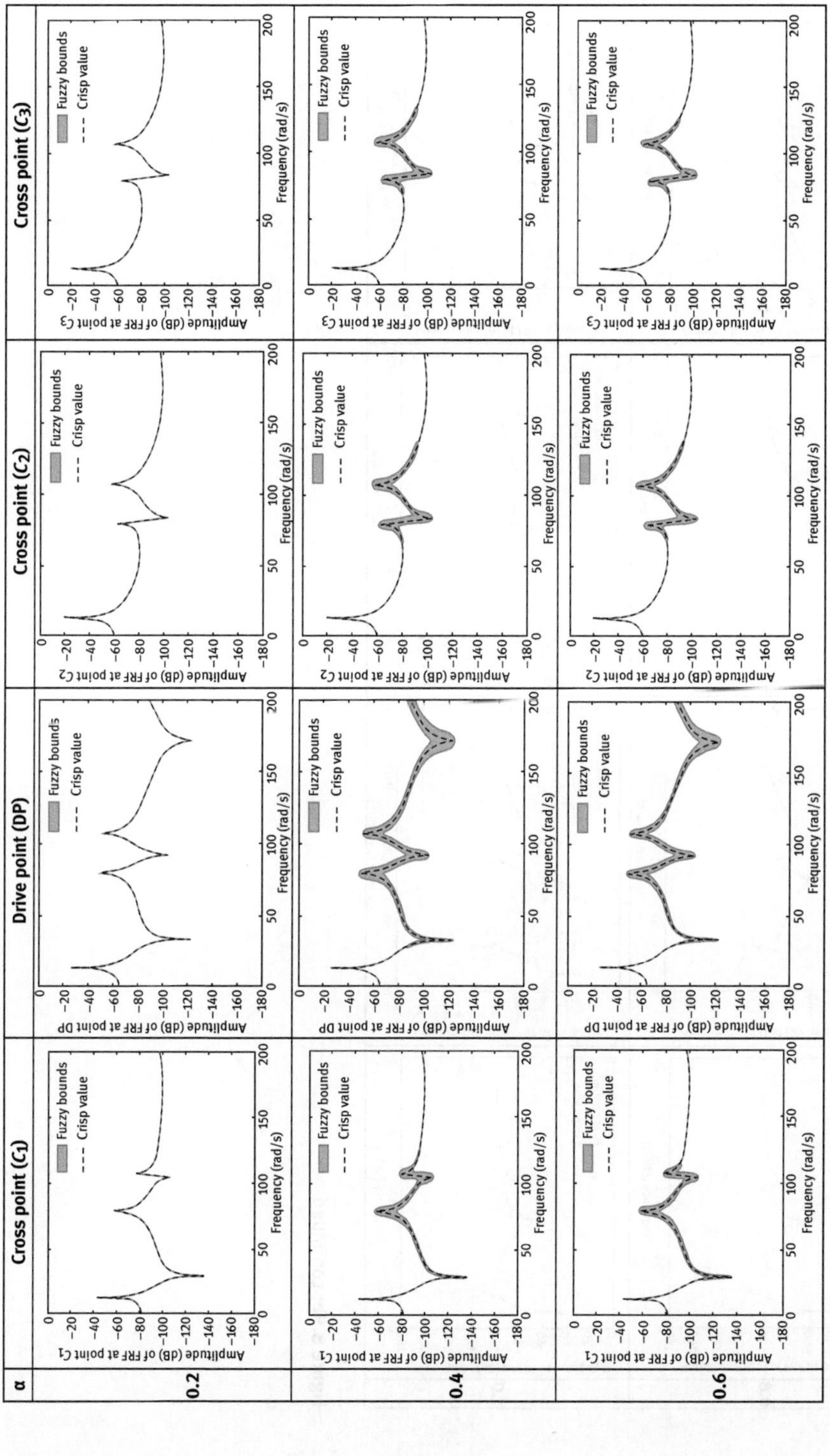

Figure 8.9: Effect of α-cut on direct fuzzy bounds and crisp values for cross points (C_1, C_2, C_3) and drive point (DP) considering combined variation of material properties of FGM cantilever plate for Figure 8.6(b).

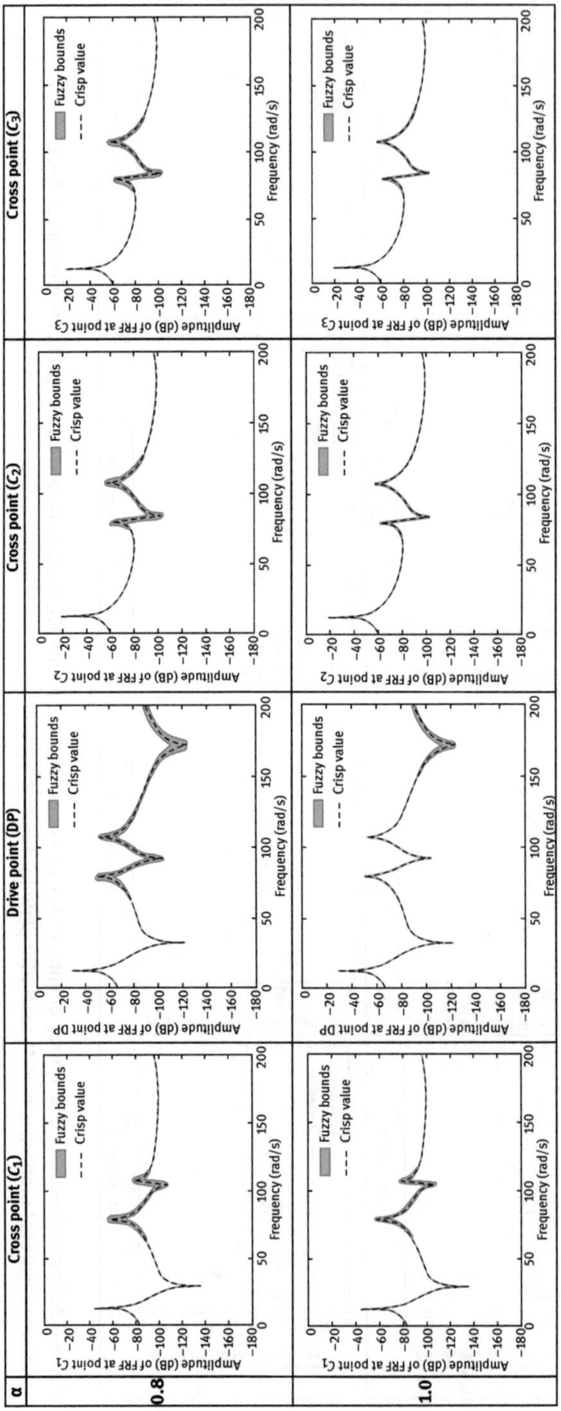

Figure 8.9b: continued

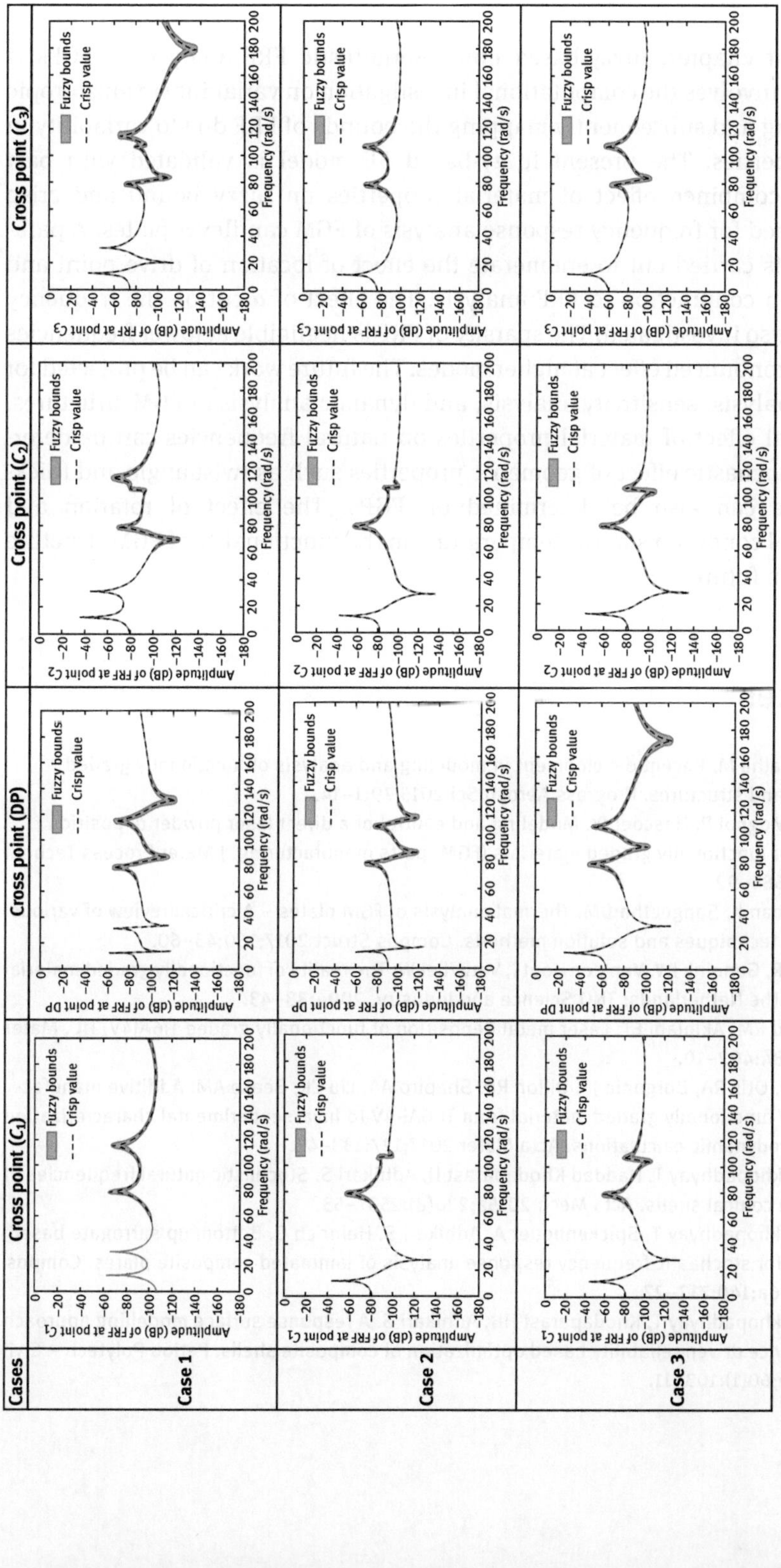

Figure 8.10: Comparative study between different locations of drive point (B) and cross points (C_1, C_2, C_3) of Figure 8.6(a), (c), and (d) on direct fuzzy bounds and crisp values for amplitude (dB)

8.6 Conclusions

In the present chapter, fuzzy-based non-deterministic FRF analysis of FGPs is carried out. It involves the computational investigation on variability in anisotropic FGM modelling and subsequently mapping the bounds of FRF due to variability in material properties. The present fuzzy-based FE model is validated with past literature. A combined effect of material properties on fuzzy bound and crisp value is mapped for frequency response analysis of FGM cantilever plates. A parametric study is carried out to enumerate the effect of location of drive point and cross points in conjunction to FRF analysis. The effect of α-cut on the frequency responses is also investigated. The sparsity of FRF is negligible at lower frequencies while it has pronounced effect at higher modes. The future work can be projected for the impact analysis, sensitivity analysis, and dynamic analysis of FGM structures. The individual effect of material properties on natural frequencies can be determined. The stochastic effect of geometric properties such as twist angle and thickness of plate can also be determined on FGPs. The effect of rotation and environmental condition such as temperature on the functionality of FGM structure can be done in future.

References

[1] Gupta A, Talha M. Recent development in modeling and analysis of functionally graded materials and structures. Progress Aerosp Sci 2015;79:1–14.
[2] Muller P, Mognol P, Hascoet JY. Modeling and control of a direct laser powder deposition process for functionally graded materials (FGM) parts manufacturing. J Mater Process Technol 2013;213;685–92.
[3] Swaminathan K, Sangeetha DM. Thermal analysis of FGM plates – A critical review of various modelling techniques and solution methods. Compos Struct 2017;160;43–60.
[4] Knoppers R, Gunnink JW, Van den Hout J, Van Vliet W. The reality of functionally graded material products. The Netherlands: TNO Science and Industry, 2004:38–43.
[5] Mahamood RM, Akinlabi ET. Laser metal deposition of functionally graded Ti6Al4V/TiC. Mater Des 2015;84:402–10.
[6] Bobbio LD, Otis RA, Borgonia JP, Dillon RP, Shapiro AA, Liu ZK, Beese AM. Additive manufacturing of a functionally graded material from Ti-6Al-4V to Invar: experimental characterization and thermodynamic calculations. Acta Mater 2017;127:133–42.
[7] Dey S, Mukhopadhyay T, Haddad Khodaparast H, Adhikari S. Stochastic natural frequencies of composite conical shells. Acta Mech 2015a;226(8):2537–53.
[8] Dey S, Mukhopadhyay T, Spickenheuer A, Adhikari S, Heinrich G. Bottom up surrogate based approach for stochastic frequency response analysis of laminated composite plates. Compos Struct 2016a;140:712–27.
[9] Dey S, Mukhopadhyay T, Khodaparast HH, Adhikari S. A response surface modelling approach for resonance driven reliability based optimization of composite shells. Period Polytech – Civil Eng 2016e;60(1):103–11.

[10] Dey S, Mukhopadhyay T, Adhikari S, Spickenheuer A, Gohs U. Uncertainty quantification in natural frequency analysis of composite plates – an Artificial neural network based approach. Adv Compos Lett 2016c;25(2):43–8.
[11] Dey S, Mukhopadhyay T, Adhikari S. Metamodel based high-fidelity stochastic analysis of composite laminates: a concise review with critical comparative assessment. Compos Struct 2017;171 227–50.
[12] Mukhopadhyay T, Naskar S, Dey S, Adhikari S. On quantifying the effect of noise in surrogate based stochastic free vibration analysis of laminated composite shallow shells. Compos Struct 2016a;140:798–805.
[13] Mukhopadhyay T, Chakraborty S, Dey S, Adhikari S, Chowdhury R. A critical assessment of Kriging model variants for high-fidelity uncertainty quantification in dynamics of composite shells. Arch Comput Methods Eng 2017a;24(3):495–518.
[14] Dey S, Mukhopadhyay T, Sahu SK, Adhikari S. Stochastic dynamic stability analysis of composite curved panels subjected to non-uniform partial edge loading. Eur J Mech A Solids 2018a;67:108–22.
[15] Karsh PK, Mukhopadhyay T, Dey S. Spatial vulnerability analysis for the first ply failure strength of composite laminates including effect of delamination. Compos Struct 2018;184:554–67.
[16] Dey S, Mukhopadhyay T, Khodaparast HH, Kerfriden P, Adhikari S. Rotational and ply-level uncertainty in response of composite shallow conical shells. Compos Struct 2015b;131 594–605.
[17] Naskar S, Mukhopadhyay T, Sriramula S, Adhikari S. Stochastic natural frequency analysis of damaged thin-walled laminated composite beams with uncertainty in micromechanical properties. Compos Struct 2017;160:312–34.
[18] Dey S, Mukhopadhyay T, Sahu SK, Adhikari S. Effect of cutout on stochastic natural frequency of composite curved panels. Composites Part B: Eng 2016d;105:188–202.
[19] Dey S, Mukhopadhyay T, Naskar S, Dey TK, Chalak HD, Adhikari S. Probabilistic characterization for dynamics and stability of laminated soft core sandwich plates. J Sandwich Struct Mater 2018b. DOI:10.1177/1099636217694229
[20] Mukhopadhyay T, Adhikari S. Equivalent in-plane elastic properties of irregular honeycombs: an analytical approach. Int J Solids Struct 2016d;91:169–84.
[21] Mukhopadhyay T, Adhikari S. Effective in-plane elastic properties of auxetic honeycombs with spatial irregularity. Mech Mater 2016e;95:204–22.
[22] Mukhopadhyay T, Adhikari S. Free vibration analysis of sandwich panels with randomly irregular honeycomb core. J Eng Mech 2016f;142(11):06016008.
[23] Mukhopadhyay T, Adhikari S, Batou A. Frequency domain homogenization for the viscoelastic properties of spatially correlated quasi-periodic lattices. Int J Mech Sci 2018. DOI:10.1016/j.ijmecsci.2017.09.004
[24] Reddy JN. A review of the literature on finite-element modeling of laminated composite plates. Shock Vib Digest 1985;17(4):3–8.
[25] Xu Y, Qian Y, Chen J, Song G. Stochastic dynamic characteristics of FGM beams with random material properties. Compos Struct 2015;133:585–94.
[26] Xu Y, Qian Y, Song G. Stochastic finite element method for free vibration characteristics of random FGM beams. Appl Math Modell 2016;40:10238–53.
[27] Hien TD, Noh HC. Stochastic isogeometric analysis of free vibration of functionally graded plates considering material randomness. Comput Methods Appl Mech Eng 2017;318:845–63.
[28] Jagtap KR, Lal A, Singh BN. Stochastic nonlinear free vibration analysis of elastically supported functionally graded materials plate with system randomness in thermal environment. Compos Struct 2011;93:3185–99.
[29] Lin RM Function-weighted frequency response function sensitivity method for analytical model updating. J Sound Vib 2017;403:59–74.

[30] Deng J, Liu Y, Zhang Z, Liu W. Stability analysis of multi-span viscoelastic functionally graded material pipes conveying fluid using a hybrid method. Eur J Mech A/Solids 2017;65:257–70.

[31] Wang Z, Zhu P. Response prediction for modified mechanical systems based on in-situ frequency response functions: theoretical and numerical studies. J Sound Vib 2017;400:417–41.

[32] de Lima AMG, Faria AW, Rade DA. Sensitivity analysis of frequency response functions of composite sandwich plates containing viscoelastic layers. Compos Struct 2010;92:364–76.

[33] Yaghoubi V, Marelli S, Sudret B, Abrahamsson T. Sparse polynomial chaos expansions of frequency response functions using stochastic frequency transformation. Probab Eng Mech 2017;48:39–58.

[34] Mao Z, Todd M. Statistical modeling of frequency response function estimation for uncertainty quantification. Mech Syst Signal Process 2013;38:333–45.

[35] Wang JT, Wang CJ, Zhao JP. Frequency response function-based model updating using Kriging model. Mech Syst Signal Process 2017;87:218–28.

[36] Li L, Zhang DG. Free vibration analysis of rotating functionally graded rectangular plates. Compos Struct 2016;136:493–504.

[37] Nguyen TK, Nguyen VH, Dinh TC, Vo TP, Xuan HN. Static and vibration analysis of isotropic and functionally graded sandwich plates using an edge-based MITC3 finite elements. Composites Part B 2016;107:162–73.

[38] Jandaghian AA, Rahmani O. Vibration analysis of functionally graded piezoelectric nanoscale plates by nonlocal elasticity theory: an analytical solution. Superlattices and Microstruct 2016;100:57–75.

[39] Gersem HD, Moens D, Desmet W, Vandepitte D. A fuzzy finite element procedure for the calculation of uncertain frequency response functions of damped structures: part 2-Numerical case studies. J Sound Vib 2005;288:463–86.

[40] Manan A, Cooper JE. Prediction of uncertain frequency response function bounds using polynomial chaos expansion. J Sound Vib 2010;329:3348–58.

[41] Zoric ND, Simonovic AM, Mitrovic ZS, Stupar SN, Obradovic AM, Lukic NS. Free vibration control of smart composite beams using particle swarm optimized self-tuning fuzzy logic controller. J Sound Vib 2014;333:5244–68.

[42] Dey S, Mukhopadhyay T, Haddad Khodaparast H, Adhikari S. Fuzzy uncertainty propagation in composites using Gram-Schmidt polynomial chaos expansion. Appl Math Modell 2016b;40(7–8): 4412–28.

[43] Moghaddam MR, Baradaran GH. Three-dimensional free vibrations analysis of functionally graded rectangular plates by the meshless local Petrov–Galerkin (MLPG) method. Appl Math Comput 2017;304:153–63.

[44] Seong HK, Yoo J. Probability distribution function inspired structural optimization for frequency response problems. Comput Methods Appl Mech Eng 2017;318:783–802.

[45] Yu C, Wang Z, Wang QJ. Analytical frequency response functions for contact of multi-layered materials. Mechanics of Materials 2014;76:102–20.

[46] Song Q, Shi J, Liu Z. Vibration analysis of functionally graded plate with a moving mass. Appl Math Modell 2017;46:141–60.

[47] Zoghaib L, Mattei P-O. Time and frequency response of structures with frequency dependent, non-proportional linear damping. J Sound Vib 2014;333:887–900.

[48] Kumar S, Mitra A, Roy H. Forced vibration response of axially functionally graded non-uniform plates considering geometric nonlinearity. Int J Mech Sci 2017;128–29:194–205.

[49] Duc ND, Lee J, Thoi TN, Thang PT. Static response and free vibration of functionally graded carbon nanotube-reinforced composite rectangular plates resting on Winkler–Pasternak elastic foundations. Aerosp Sci Technol 2017;68:391–402.

[50] Loy CT, Lam KY, Reddy JN. Vibration of functionally graded cylindrical shells. Int J Mech Sci 1999;41:309–24.
[51] Singh H, Hazarika BC, Dey S. Low velocity impact responses of functionally graded plates. Procedia Eng 2017;173:364–270.
[52] Rayleigh JW. Theory of sound. re-issue, 2nd edition. New York, Dover Publications, 1945.
[53] Bathe KJ. Finite element procedures in engineering analysis. New Delhi: PHI, 1990.
[54] Zadeh LA. Fuzzy sets. Inf Control 1965;8(3):338–53.
[55] Hanss M, Willner K. A fuzzy arithmetical approach to the solution of finite element problems with uncertain parameters. Mech Res Commun 2000;27(3):57–272.
[56] Witteveen JAS, Bijl H. Modeling arbitrary uncertainties using Gram-Schmidt polynomial chaos. AIAA paper, AIAA 2006-896. 44th AIAA Aerospace Sciences Meeting and Exhibit, Reno, Nevada, 2006.
[57] Adhikari S, Khodaparast HH. A spectral approach for fuzzy uncertainty propagation in finite element analysis. Fuzzy Sets Syst 2014;243:1–24.
[58] Hasani A. Baferani, Saidi A.R., Ehteshami H. Accurate solution for free vibration analysis of functionally graded thick rectangular plates resting on elastic foundation. Compos. Struct., 2011;93:1842–1853.
[59] Ta HD, Noh HC. Analytical solution for the dynamic response of functionally graded rectangular plates resting on elastic foundation using a refined plate theory. Appl Math Model 2015;39:6243–57.
[60] Mukhopadhyay T, Adhikari S. Stochastic mechanics of metamaterials. Compos Struct 2017b;162:85–97.
[61] Mukhopadhyay T, Adhikari S. Effective in-plane elastic moduli of quasi-random spatially irregular hexagonal lattices. Int J Eng Sci 2017d;119:142–79.
[62] Mukhopadhyay T, Mahata A, Adhikari S, Asle Zaeem M. Effective elastic properties of two dimensional multiplanar hexagonal nano-structures. 2D Mater 2017c;4 025006.
[63] Mukhopadhyay T, Mahata A, Adhikari S, Asle Zaeem M. Effective mechanical properties of multilayer nano-heterostructures. Sci Rep 2017e;7:15818.
[64] Mahata A, Mukhopadhyay T, Adhikari S. A polynomial chaos expansion based molecular dynamics study for probabilistic strength analysis of nano-twinned copper. Mater Res Express 2016;3:036501.
[65] Mukhopadhyay T, Dey TK, Dey S, Chakrabarti A. Optimization of fiber reinforced polymer web core bridge deck – A hybrid approach. Struct Eng Int 2015a;25(2):173–83.
[66] Mukhopadhyay T, Dey S, Chowdhury R, Chakrabarti A, Adhikari S. Optimum design of FRP bridge deck: an efficient RS-HDMR based approach. Struct Multidiscip Optim 2015b;52(3):459–77.
[67] Mukhopadhyay T, Mahata A, Dey S, Adhikari S. Probabilistic analysis and design of HCP nanowires: an efficient surrogate based molecular dynamics simulation approach. J Mater Sci Technol 2016b;32(12):1345–51.
[68] Mukhopadhyay T, Chowdhury R, Chakrabarti A. Structural damage identification: a random sampling-high dimensional model representation approach. Adv Struct Eng 2016c;19(6):908–27.
[69] Dey TK, Mukhopadhyay T, Chakrabarti A, Sharma UK. Efficient lightweight design of FRP bridge deck, Proc Inst Civil Eng – Struct Build 2015c;168(10):697–707.
[70] Mukhopadhyay T. A multivariate adaptive regression splines based damage identification methodology for web core composite bridges including the effect of noise. J Sandwich Struct Mater 2018. DOI:10.1177/1099636216682533

Akant Kumar Singh, Siddhartha, and Prashant Kumar Singh

9 Transmission efficiency of glass fiber-filled functionally graded material-based PA66 composite spur gears

Abstract: In the last few years, metallic gears have been replaced by polymer gears in various applications, mainly in low and moderate loading conditions. They have lower inertia, less weight and run much quieter than their metal counterparts. These characteristics are further enhanced in case of polymer composite-based gears. In the present work, the transmission efficiency of glass fiber-reinforced homogeneous polyamide 66 (PA66) composite gears and functionally graded material (FGM)-based PA66 gears has been investigated. Homogeneous gears are fabricated by conventional technique, whereas FGM-based PA66 gears are fabricated using a novel manufacturing system. Unfilled PA66 gears are also fabricated by horizontal centrifugal casting technique using injection molding machine for comparative study. PA66 pellets reinforced with 15 and 30 wt% glass fibers are used to fabricate homogeneous as well as FGM gears. Experimental work is performed on a polymer gear test rig. Experiments are performed at various speed and torque combinations for 0.2 million cycles. Results show that the operating torque has more impact on the transmission efficiency of the fabricated gears as compared to rotational speed.

Keywords: PA66, FGM gear, polymer gear test rig, transmission efficiency

9.1 Introduction

Plastic gears are found today in both commercial- and precision-type applications for transfer of motion. Transmission efficiency is one of the key performance parameters for power and motion transmission applications. The overall mechanical system efficiency in some power transmission applications has improved by using the plastic gears [1]. Plastic gears are used in various applications nowadays. However, the transmission efficiency of polymer gears has not been investigated much so far. Walton et al. [2, 3] observed that speed and loads are the key factors that significantly affect the transmission efficiency of polymer gears. They investigated the effect of different materials and geometries on the transmission efficiency of polymer gear pairs.

Polymer gears have been the focus of many researchers since the last decade [4]. Senthilvelan and Gnanamoorthy [5] investigated the efficiency of injection-molded polymer composite spur gears and found that carbon-fiber-reinforced nylon gears exhibit less temperature and better transmission efficiency than unreinforced nylon

https://doi.org/10.1515/9783110545104-009

gears due to its improved gear tooth stiffness and thermal characteristics. Deterioration that occurs at gear tooth surface during running reduces the efficiency as the cycle time increases. Transmission efficiency of polymer gears is also improved with the addition of nanoclay particles [6]. Transmission efficiency of polymer gears is improved to about 2% by using compressed air cooling at the mating surface of master and test gears [7]. Rayudu and Nagarajan [8] studied the effect of tooth deformation and gear wear on the transmission characteristics of polymer gear. Luscher et al. [9] investigated the transmission error and geometry of polyketone gears in which transmission efficiency and effect of packing pressure on gear runout is studied. Mertens and Senthilvelan [10] investigated the surface durability of injection-molded carbon nanotube–polypropylene spur gears and found that inclusion of carbon nanotube improved the transmission efficiency of polymer gears. Experimental and numerical evaluation of transmission characteristics of injection-molded polymer spur gears are done by Kodeeswaran et al. [11]

Polyamide is a promising material to fabricate polymer gears. Researchers have been using polyamide to fabricated polymer gears [12–16]. Fiber-reinforced polyamide has also been used to fabricate polymer composite gears [17–20]. Duzcukoglu [12] carried out an experimental study in which polyamide gear teeth were modified in order to distribute the generated heat on the tooth surface by means of drilled cooling holes at different locations on the gear tooth body. It resulted in reduced tooth surface temperature and led to an increase in the load-carrying capacity and improved the wear resistance of the gear teeth. Duzcukoglu [13] also improved the durability of polyamide 66 (PA66) spur gear with tooth width modification. Kodeeswaran et al. [14] investigated the effects of frequency on hysteretic heating and fatigue life of unreinforced injection-molded PA66 spur gears and observed that hysteretic heating and surface temperature increases as well as fatigue life of the gear decreases with the increase in the torque for both the bidirectional and unidirectional loads. Kodeeswaran et al. [15] also investigated the effect of strain rate on bending and transmission characteristics of injection-molded PA66 spur gears. It was observed that bending stress in the teeth increased with increasing strain rate. It happened due to higher material modulus at higher strain rates. Zemlyakov [16] investigated the effect of strain rate on the stiffness of polyamide gear teeth. The performance of plastic gear made of carbon fiber-reinforced polyamide 12 (PA12) is observed by Kurokawa et al. [17]. It was found that carbon fiber-filled PA12 gear had an excellent wear property when grease is applied at the engagement region. Gurunathan et al. [18] observed that wear resistance of polyamide nanocomposite spur gears enhanced by incorporation of nanoparticulate. Wear resistance of polymer gear teeth is also enhanced by the addition of fibers in polymer gears [19]. Cathelin et al. [20] developed a computational method to predict the mechanical behavior of plastic cylindrical gears made of fiber-reinforced polyamide 6.

The concept of functionally graded materials (FGMs) originated in Japan in 1984 [21]. It has been suggested in a review article that gear should be fabricated by FGM technique [22]. Centrifugal casting technique is generally used to fabricate thermosetting-based FGMs [23–29]. However, to the best of author's knowledge, FGM-based thermoplastic gear has not been fabricated so far. The focus of this study is to develop FGM-based glass fiber-reinforced PA66 gears and compare the transmission efficiency of FGM PA66 gears with the unfilled and homogeneous PA66 gears.

9.2 Experimental details

9.2.1 Gear fabrication

Glass fiber-reinforced PA66 granules are used to fabricate homogeneous and FGMs gears. The proportion of glass fibers in PA66 granules is 15 and 30 wt%. PA66 granules are preheated in a dryer at 100 °C for 2 h to remove the moisture before injection molding. A mold is designed with the aid of a modeling software Pro-E. The design of the mold is shown in Figure 9.1. A photographic view of the injection mold is shown in Figure 9.2 for a better understanding of this work. Mold has a rotatory component known as punch which is used for the fabrication of FGM gear.

Punch is rotated by a servomotor, and speed of the punch rotation is controlled by the servocontroller. The injection pressure of 7 MPa and barrel temperatures of 190, 200, 210 °C at three zones are used for the unfilled gear fabrication. Barrel

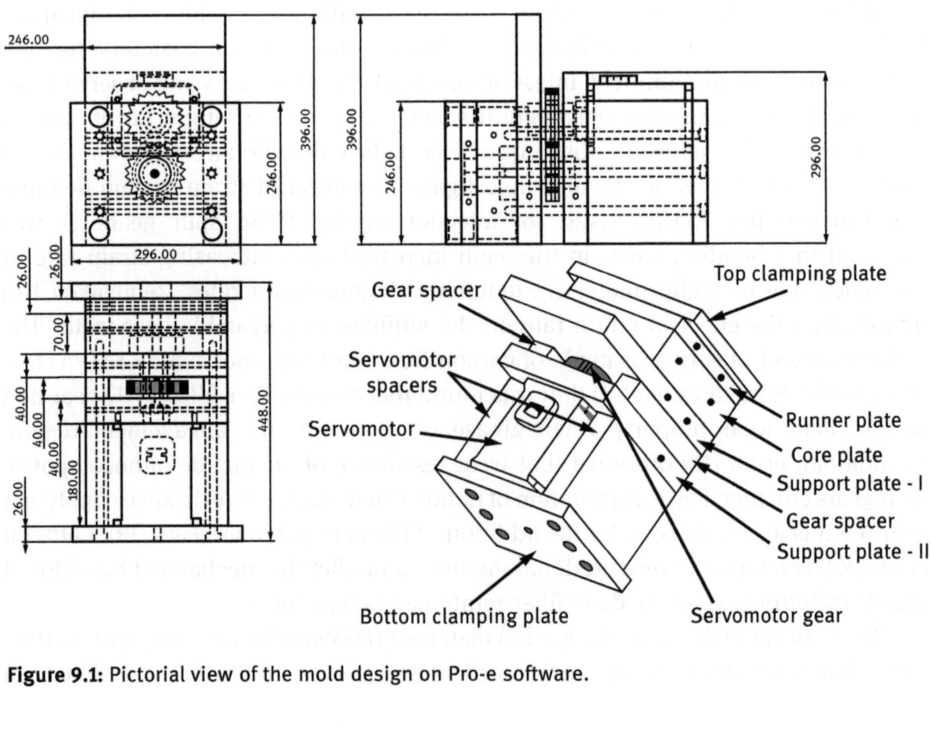

Figure 9.1: Pictorial view of the mold design on Pro-e software.

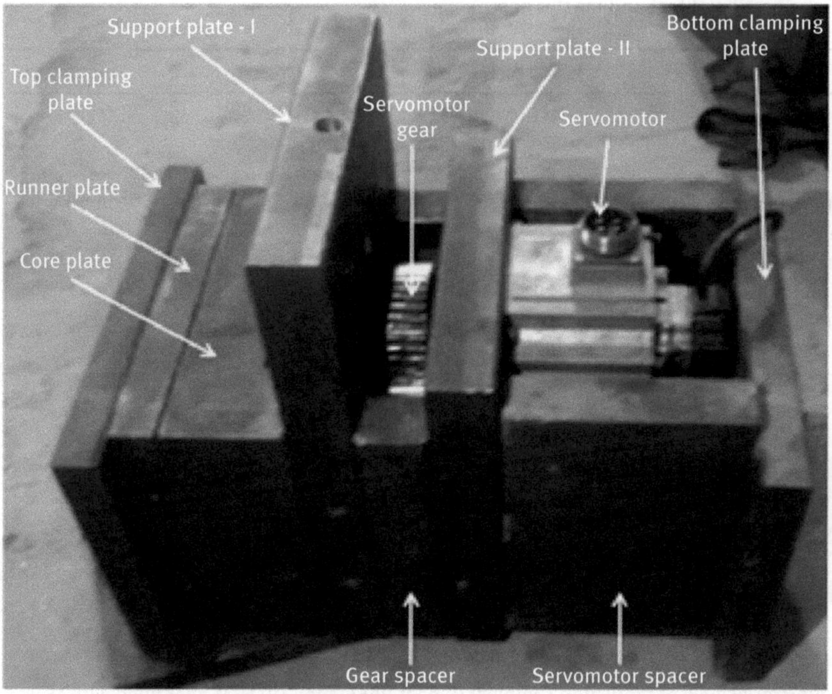

Figure 9.2: Pictorial view of injection mold.

temperatures of 210 °C, 220 °C, 230 °C and 220 °C, 230 °C, 240 °C are used for the fabrication of 15% and 30% glass fiber-filled polymer gears, respectively.

Punch is rotated at 1,800 rpm for 2 min after filling the molten material into the punch cavity to prepare FGM gear. Punch is kept stationary for homogeneous gears fabrication. Unfilled PA66 gear is also fabricated for comparing the performance with glass fiber-reinforced gears. Neat (or unfilled) and glass fiber-filled PA66 granules are supplied by Ester Industries Ltd., Uttarakhand, India. A pictorial view of fabricated gears is shown in Figure 9.3. Table 9.1 shows the specification of fabricated gears.

9.2.2 Polymer gear test rig

Gear tests are conducted using the power absorption-type gear test rig. The polymer-based test gear is driven using an AC motor to enable gear tests to be made at different speeds. The mating standard steel gear meshes with test gear to be rotated by it, but steel gear shaft is coupled to a DC motor to apply load. The principle to conduct the test is by rotating the test gear mating with standard steel gear at a specified speed and torque value for fixed number of cycles.

Table 9.1: Specification of test and master gear.

	Test gear	Master gear
Material	i. Unfilled PA66 ii. 15 wt% glass fiber-filled PA66 iii. 30 wt% glass fiber-filled PA66	Steel
Manufacturing method	Injection molded	Hobbing
Pitch circle diameter (mm)	40	40
Pressure angle (deg.)	20	20
Number of teeth	20	20
Module (mm)	2	2
Face width (mm)	8	10

(i) (ii) (iii) (iv) (v)

Figure 9.3: A pictorial view of fabricated samples: (i) neat PA66 gear, (ii) FGM gear filled with 15 wt% glass fiber, (iii) FGM gear filled with 30 wt% glass fiber, (iv) homogeneous gear filled with 15 wt% glass fiber and (v) homogeneous gear filled with 30 wt% glass fiber.

The noise generated due to the meshing of gears is captured by the acoustic sensor. The acoustic sensor is positioned near the mating portion of gears. The vibration experienced by gear shaft while rotation is measured by an accelerometer mounted on plumber block. The arrangement of acoustic sensor and accelerometer in polymer gear test rig is shown in Figure 9.4. Test gears are operated up to 0.2 million cycles in the present investigation.

9.2.3 Methodology to verify the gradation in FGM specimen

Ignition loss test of homogeneous and FGM gears is performed as per ASTM D2584 standard. Samples are prepared for the ignition loss test by cutting a piece from the gear. This piece is further cut into three zones as shown in Figure 9.5.

Ignition loss test is carried out by burning the cut samples in electric muffle furnace at 565 ± 28 °C using a crucible and weighing the remaining glass fibers. In ignition loss test, the polymer matrix material is burned out completely and the remaining glass fibers are collected in the crucible. Three samples for each zone of homogeneous and FGM gear are used for ignition loss test and the average of three

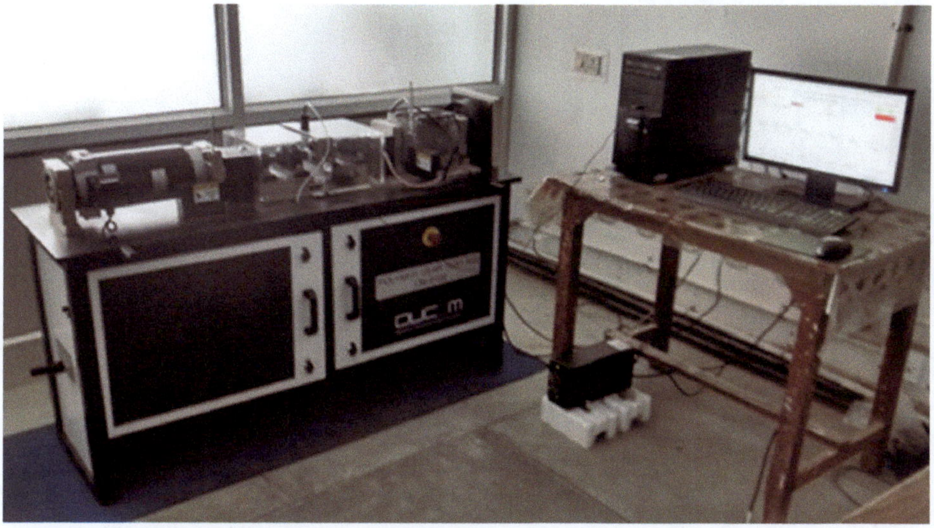

Figure 9.4: A pictorial view of polymer gear test rig.

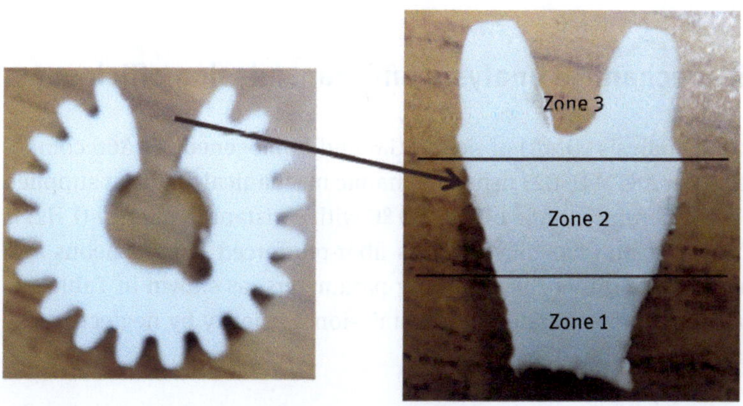

Figure 9.5: Zone of the gear for ignition loss test.

values is taken into the study. Ignition loss of homogeneous and FGM gear is calculated in weight percentage as follows:

$$\text{Ignition loss, weight \%} = [(W_1 - W_2)/W_1] \times 100 \tag{9.1}$$

where W_1 is the weight of the sample, g; W_2 is the weight of residue, g.

A digital electronic balance (model: TB-215D supplied by Denver Instrument) with an accuracy of 0.00001 g is used to measure the weight of homogeneous and FGM specimens. The hardness of neat PA66, homogeneous and FGM gears is

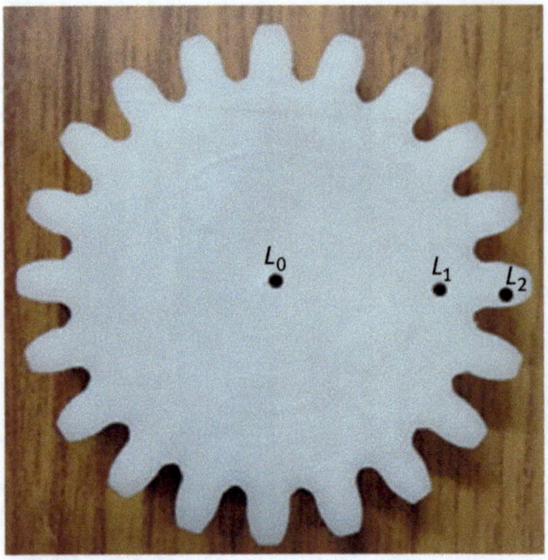

Figure 9.6: Representation of locations for hardness test.

measured using a digital durometer as per ASTM D2240 standard. Hardness is measured at three locations as shown in Figure 9.6. Hardness is measured five times at each location and average of five values is taken in the study.

9.2.4 Dynamic mechanical analysis and transmission efficiency

Dynamic mechanical analysis (DMA) of neat, FGM and homogeneous PA66 composites is performed as per ASTM D5023 using a dynamic mechanical analyzer supplied by Perkin Elmer. A temperature range of 20–140 °C with constant frequency (1 Hz) is used for DMA. Transmission efficiency of glass fiber-reinforced homogeneous and FGM gears are investigated under the operating parameters, as shown in Table 9.2. Following relation is used to evaluate the transmission efficiency by neglecting the power loss at couplings and bearings:

$$\text{Transmission efficiency} = \frac{\text{Driven gear torque}}{\text{Driver gear torque}} \tag{9.2}$$

Table 9.2: Experimental parameters of homogeneous and FGM gears.

Gear designation	Gear material	Torque (Nm)	Speed (rpm)
NPA66G	Neat PA66		
PA66H15G	PA66 + 15 wt% glass fiber-reinforced homogeneous gear	0.8	500
PA66H30G	PA66 + 30 wt% glass fiber-reinforced homogeneous gear	1.4	800
PA66F15G	PA66 + 15wt% glass fiber-reinforced FGM gear	2.0	1,100
PA66F30G	PA66 + 30wt% glass fiber-reinforced FGM gear	2.6	1,400

9.3 Results and discussions

9.3.1 Experimental verification of gradation in FGM gears by ignition loss test

Ignition loss of FGM and homogeneous gear in weight percentage is shown in Figure 9.7. It is observed that 15 and 30 wt% glass fiber-filled homogeneous gears have almost similar amount of ignition loss at each zone. It reflects an even distribution of glass fibers in homogeneous gears. A reasonable reduction in ignition loss is observed in FGM gears from zone 1 to zone 3 due to the gradation of the fibers toward gear tooth as evident from Figure 9.7. Homogeneous gears have less ignition loss as compared to FGM gears in zone 1 as observed from Figure 9.7. It happens because zone 1 in FGM gears become matrix-rich zone due to the gradation of the fibers toward gear tooth (i.e., zone 3).

9.3.2 Experimental verification of gradation in FGM gears by hardness measurement

Hardness measurement is done to verify the gradation of fibers in FGM gears. Hardness is measured along the radius of the FGM gear at three different locations from the center of gear to the gear tooth. Hardness values at three different locations are shown in Table 9.3. Hardness is measured five times at each location and average of five values is taken into the study as listed in Table 9.3. It is observed from Table 9.3 that hardness increases from the center of FGM gear toward the gear tooth. The

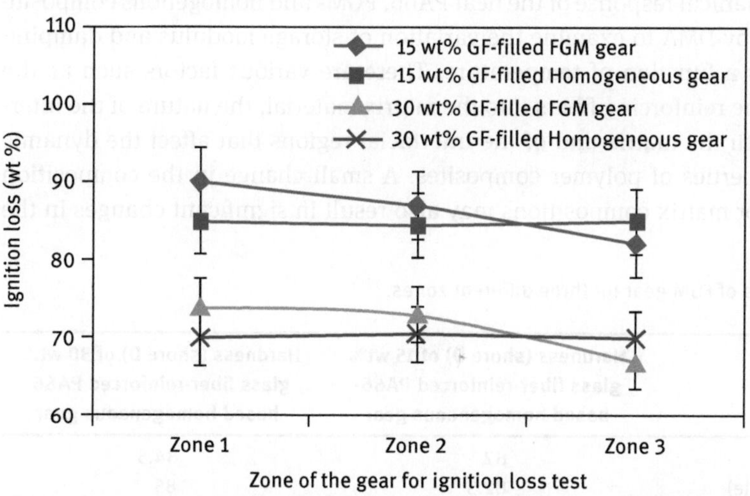

Figure 9.7: Representation of zones for ignition loss test.

Table 9.3: Hardness of FGM gear for three different zones.

Location	Hardness (shore D) of 15 wt% glass fiber-reinforced PA66-based FGM gear	Hardness (shore D) of 30 wt% glass fiber-reinforced PA66-based FGM gear
L_0 (at the center)	81	83
L_2 (near to root circle)	84	86.5
L_3 (in the gear tooth)	86	88

hardness of 30 wt% glass fiber-reinforced FGM gear is found to be higher as compared to 15 wt% glass fiber-reinforced FGM gear in all three zones as shown in Table 9.3.

Hardness is also measured for homogeneous gears at same three locations as FGM gears. The hardness of homogeneous gears is shown in Table 9.4. Variation in the hardness values is only (0–1) Shore D for homogeneous gears in all three locations that indicates an even distribution of glass fibers in homogeneous gears.

Homogeneous gear reinforced with 30 wt% glass fiber exhibits higher hardness as compared to 15 wt% glass fiber-reinforced homogeneous gear. Hardness measured at three different locations of FGM and homogeneous gears is shown in Figure 9.6. Neat PA66 gear exhibits a hardness of 77 shore D. Unfilled gear shows minimum hardness while 30 wt% glass fiber-reinforced FGM gear has maximum hardness among all the fabricated gears.

9.3.3 Dynamic mechanical analysis

The thermomechanical response of the neat PA66, FGMs and homogenous composite gears is studied by DMA to examine the variation of storage modulus and damping factor (tan δ) as a function of temperature. There are various factors such as the distribution of the reinforcing fibers into the matrix material, the nature of the interface of fibers with the matrix and of the interfacial regions that affect the dynamic mechanical properties of polymer composites. A small change in the composition regarding fiber or matrix compositions may also result in significant changes in the

Table 9.4: Hardness of FGM gear for three different zones.

Location	Hardness (shore D) of 15 wt% glass fiber-reinforced PA66-based homogeneous gear	Hardness (shore D) of 30 wt.% glass fiber-reinforced PA66-based homogeneous gear
L_0 (at the center)	82	84.5
L_2 (near to root circle)	82.5	85
L_3 (in the gear tooth)	83	84.5

overall dynamic mechanical properties of the composite [30]. Damping behavior of composites can be used to predict temperatures, stiffness and impact properties. Figures 9.8 and 9.9 represent the storage modulus and damping factor of neat PA66, glass fiber-filled PA66-based FGMs and homogeneous composites, respectively.

The stiffness of a viscoelastic material is represented through the storage modulus. The material shows stable mechanical behavior at the temperature less than 40 °C as the composites are in the glassy region and thus there is limited molecular mobility of the polymeric chains. The storage modulus of all composites steeply decreases with further increase in temperature for the range of 40–85 °C as evident from Figure 9.8. The order of storage modulus in the temperature range of 40–85 °C is PA66F30 > PA66H30 > PA66F15 > PA66H15 > NPA66, which indicates reinforcement effects divulged by the fiber allowing stress transfer ability across the interface.

The damping factor (tan δ) indicates the recoverable energy in terms of mechanical damping or internal friction in a viscoelastic system [31]. The peak of the damping factor (tan δ) shows the glass transition temperature of the composites. Under the glass transition stage, composite material property changes from a glass state to rubber-elastic state when the sinusoidal load at the frequency of 1 Hz is applied with the increase in temperature. FGMs reinforced with 15 and 30 wt% glass fibers have the glass transition temperature of 85 °C and 77 °C, respectively, while the homogenous composites have 85 °C and 79 °C. The glass transition temperature for neat PA66 is 70 °C. Neat PA66 and FGM filled with 30 wt% glass fiber has maximum and minimum damping factors (tan δ), respectively.

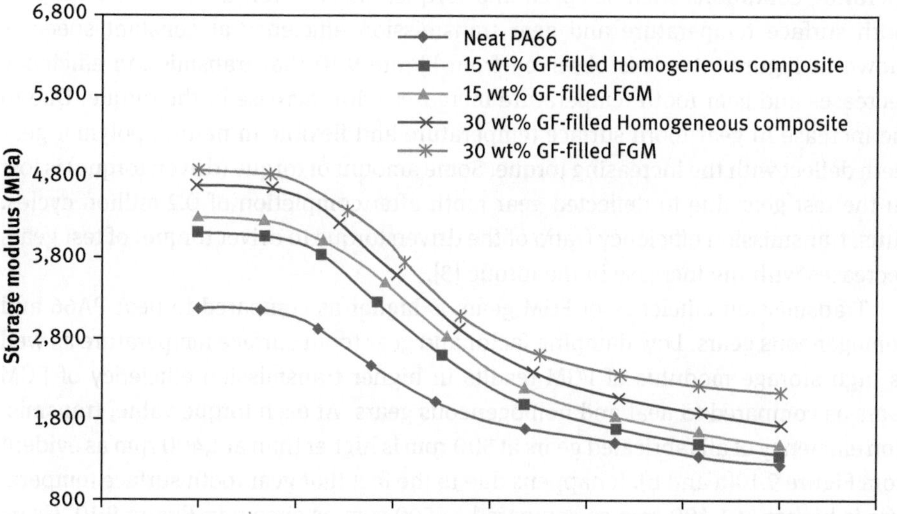

Figure 9.8: Storage modulus of PP composites.

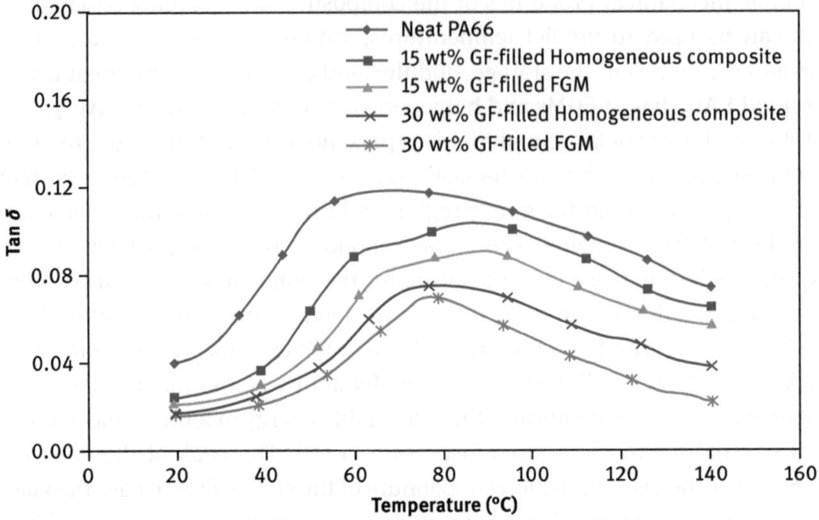

Figure 9.9: Damping factor of PP composites.

9.3.4 Effect of speed and torque on the transmission efficiency of PA66-based homogeneous and FGM gears

The transmission efficiency of the polymer gears depends on the gear material and operating conditions such as speed and torque. Effect of torque increment on gear tooth surface temperature and gear transmission efficiency at constant speed is shown in Figure 9.10. It is observed from Figure 9.10 that transmission efficiency decreases and gear tooth temperature increases with increase in the torque. Due to the increase in gear tooth surface temperature and flexible in nature, polymer gear teeth deflect with the increasing torque. Some amount of torque (driven torque) is lost on the test gear due to deflected gear tooth after completion of 0.2 million cycles. Thus, transmission efficiency (ratio of the driven torque to driver torque) of test gears decreases with the increase in the torque [3].

Transmission efficiency of FGM gears is higher as compared to neat PA66 and homogeneous gears. Low damping factor and gear tooth surface temperature as well as high storage modulus of FGM results in higher transmission efficiency of FGM gears as compared to neat and homogeneous gears. At each torque value, transmission efficiency of all fabricated gears at 500 rpm is higher than at 1,400 rpm as evident from Figure 9.10(a and b). It happens due to the fact that gear tooth surface temperature is higher at 1,400 rpm as compared to 500 rpm as shown in Figure 9.10. Gears filled with 30 wt% glass fiber have higher transmission efficiency than 15 wt% glass fiber-filled gears. Neat PA66 gear has minimum transmission efficiency at 500 rpm as well as at 1,400 rpm for each torque value. Transmission efficiency of fabricated gears

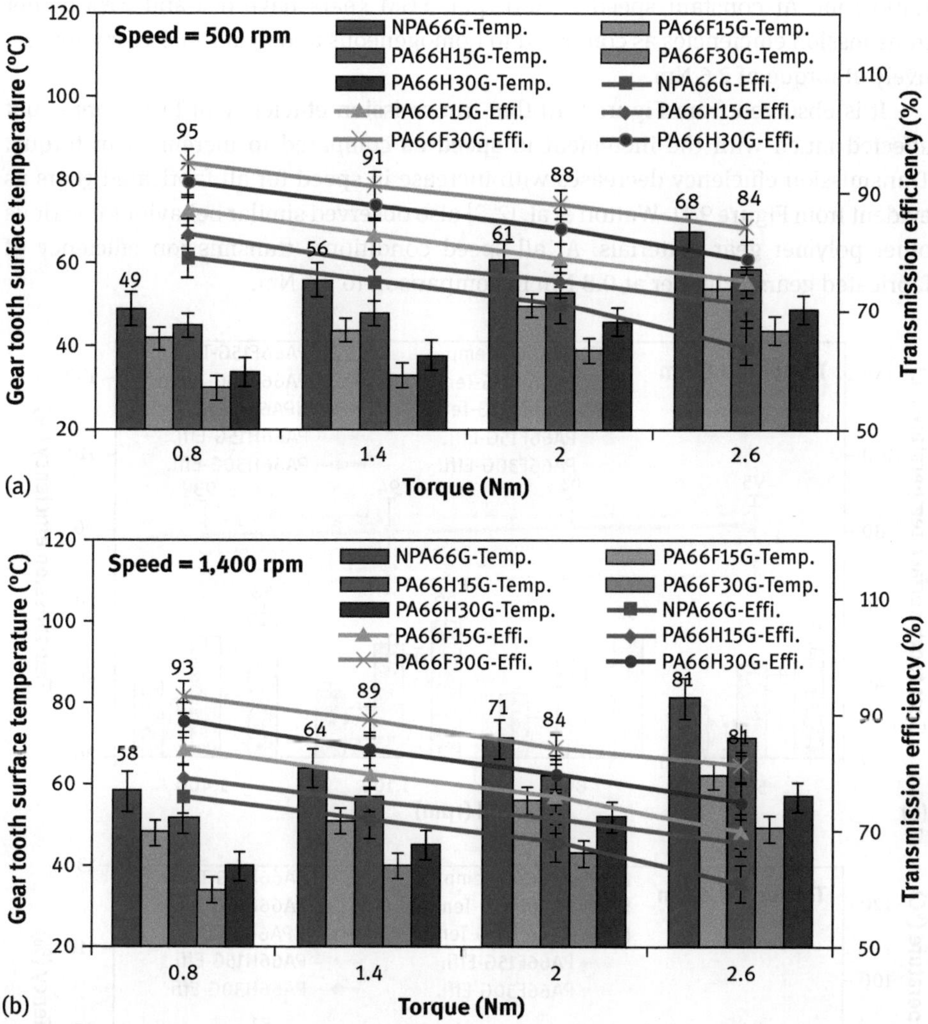

Figure 9.10: (a) Transmission efficiency of the gears at a speed of 500 rpm after 2×10^5 cycle. (b) Transmission efficiency of the gears at speed = 1,400 rpm after 2×10^5 cycle.

decreases by 19% for neat PA66 gear, 16% for homogeneous (filled with 30 wt% glass fiber) gear and 13% for FGM (filled with 30 wt% glass fiber) gear with increase in the torque from 0.8 to 2.6 Nm. It is evident that the reduction in transmission efficiency is minimum for FGM gear filled with 30 wt% glass fibers.

It happens due to the fact that FGM gear teeth have higher amount of fibers that insure higher rate of heat removal resulting in better transmission efficiency. At torque of 2.6 Nm, FGM gears have 8% and 24% higher transmission efficiency as compared to homogeneous and neat PA66 gears, respectively, at constant speed of

1,400 rpm. At constant speed of 500 rpm, FGM gears have 6% and 23% higher transmission efficiencies as compared to homogeneous and neat PA66 gears, respectively at torque of 2.6 Nm.

It is observed from Figure 9.10 that transmission efficiency of test gears is not affected much with the increment in speed as compared to increment in torque. Transmission efficiency decreases with increase in speed for all fabricated gears as evident from Figure 9.11. Walton et al. [2, 3] also observed similar behavior for various other polymer gear materials. At all speed conditions, transmission efficiency of fabricated gears is higher at 0.8 Nm in comparison to 2.6 Nm.

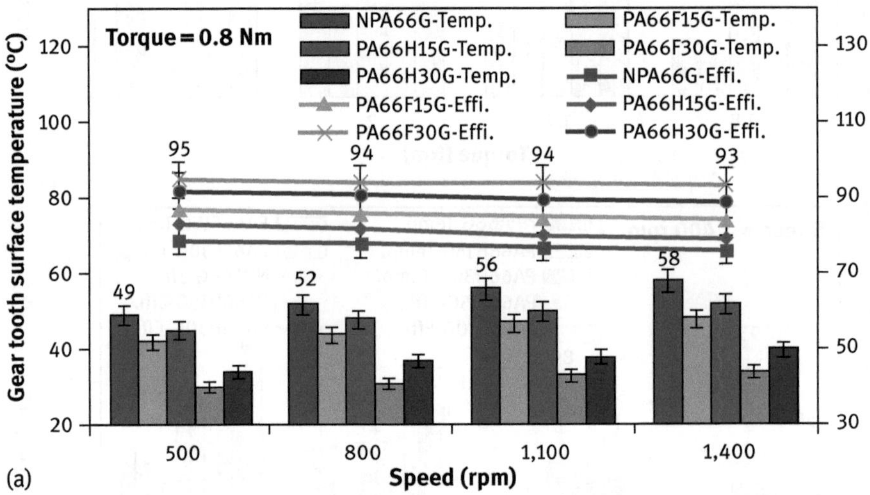

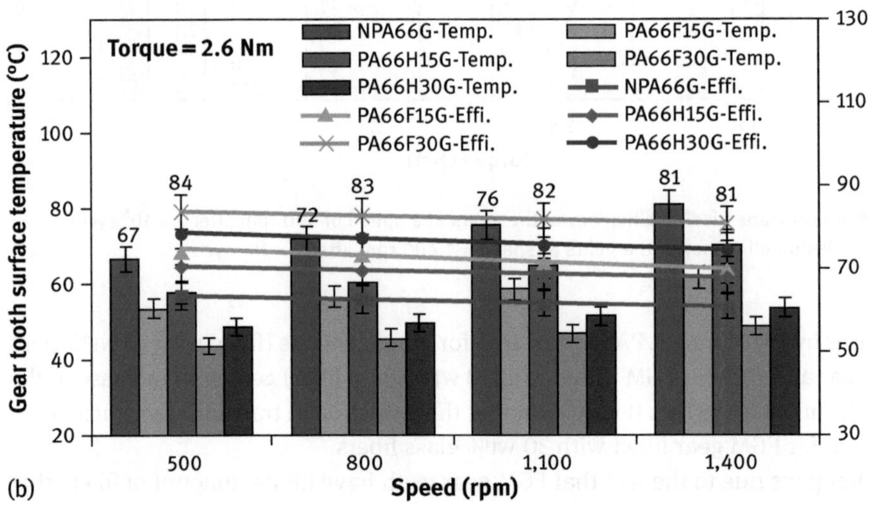

Figure 9.11: (a) Transmission efficiency of the gears at torque = 0.8 Nm after 2×10^5 cycle. (b) Transmission efficiency of the gears at torque = 2.6 Nm after 2×10^5 cycle.

It happens as the gear tooth surface temperature is higher at 2.6 Nm (Figure 9.11(b)) as compared to 0.8 Nm (Figure 11.11(a)) for all fabricated gears. Transmission efficiency of fabricated gears decreases marginally from 3% to 5% with increase in the speed from 500 to 1,400 rpm.

At constant torque of 0.8 Nm, FGM gear reinforced with 30 wt% of glass fiber has 3% and 17% higher transmission efficiency as compared to homogeneous and neat PA66 gears, respectively, at a speed of 500 rpm. At the same speed and with same weight percentage of glass fiber reinforcement, FGM gear has 6% and 24% higher transmission efficiencies as compared to homogeneous and neat PA66 gears, respectively, at constant torque of 2.6 Nm. Hence, it is observed that FGM gears perform better than the homogeneous and neat PA66 gears.

9.4 Conclusions

The following conclusions can be drawn from this research work:

1. A novel manufacturing route is developed with successful demonstration of in situ manufacturing of thermoplastic composite-based FGM gears.
2. The gradation of fibers in FGM gears is successfully verified by ignition loss test and hardness measurement.
3. Transmission efficiency of fabricated gears is much affected by torque rather than speed.
4. FGM gears have higher transmission efficiency as compared to neat and homogeneous PA66 gears.
5. Neat PA66 gear has minimum transmission efficiency among all the fabricated gears.

Acknowledgments: The authors thankfully acknowledge the support granted by the Department of Science and Technology (DST), Government of India; for this work vide their grant no.: SERB/F/5361/2012-2013.

References

[1] Adams CE. Plastic gearing: selection and application. New York: Marcel Dekker, 1986.

[2] Walton D, Cropper AB, Weale DJ, Meuleman PK. The efficiency and friction of plastic cylindrical gears. Part 1: influence of materials. P I Mech Eng J: J Eng Tribol 2002;216:75–92.

[3] Walton D, Cropper AB, Weale DJ, Meuleman PK. The efficiency and friction of plastic cylindrical gears. Part 2: influence of tooth geometry. P I Mech Eng J: J Eng Tribol 2002;216:93–103.

[4] Singh AK, Siddhartha, Singh PK. Polymer spur gears behaviors under different loading conditions: a review. P I Mech Eng J: J Eng Tribol 2017. DOI:10.1177/1350650117711595

[5] Senthilvelan S, Gnanamoorthy R. Efficiency of injection-moulded polymer composite spur gears. P I Mech Eng J: J Eng Tribol 2009;223:925–8.

[6] Kirupasankar S, Gurunathan C, Gnanamoorthy R. Transmission efficiency of polyamide nanocomposite spur gears. Mater Des 2012;39:338–43.
[7] Mertens AJ, Senthilvelan S. Durability enhancement of polymer gear using compressed air cooling. P I Mech Eng L: J Mat Des Appl 2015;230:515–25.
[8] Nagarajan T, Rayudu GV. Investigation of the transmission characteristics of plastic involute fine mechanism gears. Proceedings of the 7th World Congress on the Theory of machines and mechanisms. Sevilla, Spain, 1987:1335–8.
[9] Luscher A, Houser D, Snow C. An investigation of the geometry and transmission error of injection molded gears. J Injection Molding Technol 2000;4:177–90.
[10] Mertens AJ, Senthilvelan S. Surface durability of injection-moulded carbon nanotube–polypropylene spur gears. P I Mech Eng L: J Mat Des Appl 2016. DOI:10.1177/1464420716654308
[11] Kodeeswaran M, Suresh R, Senthilvelan S. Transmission characteristics of injection moulded polymer spur gears: experimental and numerical evaluation. Int J Power Trains 2016;5: 246–63.
[12] Duzcukoglu H. Study on development of polyamide gears for improvement of load-carrying capacity. Tribol Int 2009;42:1146–53.
[13] Duzcukoglu H. PA 66 spur gear durability improvement with tooth width modification. Mater Des 2009;30:1060–7.
[14] Kodeeswaran M, Verma A, Suresh R, Senthilvelan S. Effects of frequency on hysteretic heating and fatigue life of unreinforced injection molded polyamide 66 spur gears. Proc I Mech Eng L: J Mater Des Appl 2017. DOI:10.1177/1464420717702176
[15] Kodeeswaran M, Suresh R, Senthilvelan S. Effect of strain rate on bending and transmission characteristics of injection molded polyamide 66 spur gears. Proc IMechE Part L: J Mater: Des Appl 2017. DOI:10.1177/1464420717724484
[16] Zemlyakov IP. Effect of strain rate on the stiffness of polyamide gear teeth. Mekanika Polimerov 1996;2(3):467–9.
[17] Kurokawa M, Uchiyama Y, Iwai T, Nagai S. Performance of plastic gear made of carbon fiber reinforced polyamide 12. Wear 2003;254:468–73.
[18] Gurunathan C, Kirupasankar S, Gnanamoorthy R. Wear characteristics of polyamide nanocomposite spur gears. Proc IMechE Part J: J. Eng Tribol 2011;225:299–306.
[19] Dighe AD, Mishra AK, Wakchaure VD. Investigation of wear resistance and torque transmission capacity of glass filled polyamide and PEEK composite spur gears. Int J Eng Adv Technol 2014;3:299–303.
[20] Cathelin J, Guingand M, Vaujany JP, Chazeau L, Adrien J. Quasistatic load sharing model in the case of molded glass fiber reinforced polyamide 6 gears. Appl Compos Mater (2014). DOI:10.1007/s10443-014-9410-7.
[21] Koizumi M. FGM activities in Japan. Compos B 1997;28:1–4.
[22] Siddhartha, Patnaik A, Satapathy A, Bhatt AD. A study on modified mechanical and wear characteristics of epoxy-particulate filled homogenous composites and their functionally graded materials. J Tribol 2011;133:011601–11.
[23] Siddhartha, Patnaik A, Satapathy A, Bhatt AD. A study on modified mechanical and wear characteristics of epoxy-particulate filled homogenous composites and their functionally graded materials. J Tribol 2011;133:011601–11.
[24] Siddhartha, Patnaik A, Bhatt AD. Friction and wear analysis of a cement kiln dust reinforced epoxy-based functionally graded materials. P I Mech Eng J: J Eng Tribol 2010;224:1103–14.
[25] Siddhartha, Patnaik A, Bhatt AD. Mechanical and dry sliding wear characterization of epoxy–TiO2 particulate filled functionally graded composites materials using Taguchi design of experiment. Maters Des 2011;32:615–27.

[26] Siddhartha, Singh AK. Mechanical and dry sliding wear characterization of short glass fiber reinforced polyester-based homogeneous and their functionally graded composite materials. P I Mech Eng L: J Mat Des Appl 2015;229:274–98.
[27] Siddhartha, Singh AK, Yadav S. Exploring the possibility of utilization of red mud epoxy based functionally graded materials as wear-resistant materials using Taguchi design of experiment. Adv Polym Technol 2015;36:1–18.
[28] Singh AK, Siddhartha, Yadav S, Singh PK. Repercussion of Manufacturing Techniques on Mechanical and Wear Peculiarity of ZnO Particulate-Filled Polyester Composites. Polym Compos, (2016). DOI 10.1002/pc.23982
[29] Singh AK, Siddhartha. Mechanical and tribological peculiarity of nano-TiO_2-augmented, polyester-based homogeneous nanocomposites and their functionally graded materials. Adv Polym Technol 2016. DOI 10.1002/adv.21710
[30] Joseph PV, Mathew G, Joseph K, Groeninckx G, Thomas S. Dynamic Mechanical Properties of Short Sisal Fibre Reinforced Polypropylene Composites. Compos A 2003;34:275–90.
[31] Kuma S, Satapathy BK, Patnaik A. Thermo-mechanical correlations to erosion performance of short carbon fibre reinforced vinyl ester resin composites. Mater Des 2011;32:2260–8.

Lakshman Sondhi, Subhashis Sanyal, Kashinath Saha, and Shubhankar Bhowmick

10 Approximate solution of functionally graded thick cylinders

Abstract: The present work employs variational principle to investigate the stress and deformation states for a thick cylinder made of functionally graded material (FGM). Based on Galerkin's error minimization principle upon a series approximation, the solution of the governing equation is obtained. In the present study, the property of FGMs is based on exponential spatial variation. Stress and displacement along the cylinder radius are studied and plotted. The results are validated with benchmark and are found to be in good agreement. The results report the significance of FGMs on the cylinder stress distribution.

Keywords: functionally graded materials, stress analysis, thick cylinder.

10.1 Introduction and literature review

Continuous spatial variation of volume fraction of a mix of material results in functionally graded materials (FGMs). FGMs have evolved strongly in industrial applications in last few decades. In terms of mechanical structures, hollow cylinders are typically used as vessels for storing fluids and pipes fluid flow and have received considerable attention among the researchers. The analytical solution for displacement and stresses based on infinitesimal theory of elasticity in functionally graded (FG) internally pressurized spherical and cylindrical vessels is presented by Tutuncu and Orztuk [1]. In case of multilayered and cylindrically anisotropic pressure vessels, the stresses are reported by Verijenko et al. [2]. The thermomechanical analysis of FG cylinders under different loading conditions is reported by Tarn [3]. One-dimensional study of thermal stresses under steady state in a thick annulus made of FGM is presented by Jabbari et al. [4]. Based on multilayered approach of theory of laminated composites, the displacement and stresses for FG cylinder in thermal environment are reported by Shao [5]. Based on Lame's solution and assuming property variation along the different layers, the exact solutions of multilayered cylinders are reported by Xiang et al. [6]. Based on infinitesimal theory of elasticity, Tutuncu [7] reported power series solutions of stresses and displacement for internally pressurized FG cylindrical vessels. Closed-form solutions of stresses and strains within elastic limit in disks with geometry and density variation under thermal load is presented by Vullo and Vivio [8]. Peng and Lee [9] proposed a new technique for calculating thermal stresses in a FG thick annulus under steady state conditions. In case of press-fit orthotropic cylinders, analytical solution of the macro mechanical model

https://doi.org/10.1515/9783110545104-010

is presented by Croccolo and De Agostinis [10]. Chen and Lin [11] assumed exponential function form to determine the stresses within elastic range for FG thick annulus and reported the effect of grading on cylinder stresses.

Numerical approximation of elastic behavior of an FG hollow cylinder is reported, based on Galerkin's error minimization principle, assuming radial displacement as the unknown dependent variable. The stress and displacement field in the cylinder is evaluated based on series approximation. The relevant results are reported in dimensionless form, and hence one can readily obtain the corresponding dimensional value for all possible geometries through appropriate normalizing parameters.

10.2 Mathematical formulation

Thick cylinder of inner radius, "a", aspect ratio, "a/b" and length "h_0" having graded material property is investigated under plane strain assumption (Figure 10.1). The cylinder is investigated under internal pressure, p_a. At a certain pressure, i.e., limit pressure, the maximum equivalent stress of the cylinder exceeds the yield limit value, giving rise to yield initiation. The analysis is carried out under the assumptions of linear-elastic material behavior. Application of pressure at the inner boundary induces radial displacements depending on the boundary conditions. Based on variational principle, the solution for the displacement field is obtained using Galerkin's error minimization method. For a cylinder under plain strain assumptions, strain energy U is given as

$$U = \frac{1}{2} \int_{\text{Vol}} (\sigma\varepsilon) dv = \frac{1}{2} \int_{\text{Vol}} (\sigma_\theta \varepsilon_\theta + \sigma_r \varepsilon_r) dv \tag{10.1}$$

While external work V is given by

$$V = -(2\pi ahp) u|_a . \tag{10.2}$$

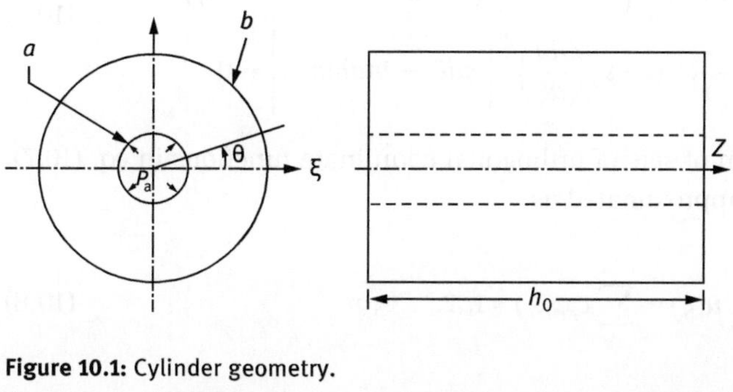

Figure 10.1: Cylinder geometry.

In the present work, $\mu = 0.3$. Exponential grading of Young's modulus [11] is considered as

$$E(r) = E_o e^{(\beta(r-a)/(b-a))} \tag{10.3}$$

E_o is the Young's modulus at inner radius and β assumes values (in the present study, −2.0 to 2.0) to represent the property of FGM and is called the gradient parameter. On substituting the constitutive relations followed by compatibility equations in eq. (10.1), strain energy is obtained as

$$U = \frac{\pi h}{(1-2\mu)(1+\mu)} \int_a^b \left[(1-\mu)\left(E(r)\frac{u^2}{r} + E(r)r\left(\frac{du}{dr}\right)^2 \right) + 2\mu E(r)(u)\left(\frac{du}{dr}\right) \right] dr \tag{10.4}$$

The solution is obtained for radial displacement field using variational method based on minimum potential energy principle:

$$\delta(U+V) = 0, \tag{10.5}$$

Substituting eqs (10.2) and (10.4) in eq. (10.5)

$$\delta \left[\frac{\pi}{(1-2\mu)(1+\mu)} \int_a^b \left[(1-\mu)\left(E(r)\frac{u^2}{r} + E(r)\frac{2\mu(u)}{(1-\mu)}\left(\frac{du}{dr}\right) + rE(r)\left(\frac{du}{dr}\right)^2 \right) \right] dr - \int_a^b up2\pi r dr \right] = 0 \tag{10.6}$$

In terms of normalized coordinate (ξ) and nondimensional space, $\xi = \frac{(r-a)}{(b-a)}$, the governing equation takes the following form:

$$\delta \left[\frac{\pi \bar{r}}{(1-2\mu)(1+\mu)} \int_0^1 \left\{ E(r)(1-\mu)\frac{u^2}{(\bar{r}\xi + a)} + E(r)\frac{2\mu\, u}{\bar{r}}\left(\frac{du}{d\xi}\right) + E(r)\frac{(\bar{r}\xi + a)}{\bar{r}^2}(1-\mu)\left(\frac{du}{d\xi}\right)^2 \right\} hd\xi - 2\pi ahpu|_0 \right] = 0 \tag{10.7}$$

Using a linear combination of sets of orthogonal coordinate functions in eq. (10.7), displacement $u(\xi)$ can be approximated as

$$u(\xi) = \sum c_j \varphi_j, \; j = 1, 2, \ldots, n \tag{10.8}$$

In eq. (10.8), φ_j are derived using Gram–Schmidt scheme wherein φ_0, start function, is selected to satisfy the boundary conditions. The boundary conditions for hollow cylinder under internal pressure p_a is as follows:

$$\sigma_r|_{(r=a)} = -p_a$$

$$\sigma_r|_{(r=b)} = 0$$

$$\varphi_0(r) = \frac{pa^2(1+\mu)}{E(r)(b^2-a^2)}\left[\left(\frac{b^2}{r}\right) + (1-2\mu)r\right] \tag{10.9}$$

The governing equation is obtained as a set of algebraic equations by substituting eq. (10.8) into eq. (10.7):

$$\delta\left[\left(\frac{\bar{r}}{(1-2\mu)(1+\mu)}\right)\int_0^1 \left\{\begin{array}{l} E(r)(1-\mu)\frac{(\Sigma c_j\varphi_j)^2}{(\bar{r}\xi+a)} + E(r)\frac{2\mu}{\bar{r}}\left[(\Sigma c_j\varphi_j)\frac{d}{d\xi}(\Sigma c_j\varphi_j)\right] \\ +E(r)\frac{(\bar{r}\xi+a)(1-\mu)}{\bar{r}^2}\left[\frac{d}{d\xi}(\Sigma c_j\varphi_j)\right]^2 \end{array}\right\} hd\xi \right.$$

$$\left. -2ahpc_j\varphi_j\Big|_0\right] = 0 \tag{10.10}$$

In eq. (10.10), the operator "δ" is replaced by $\partial/\partial c_i$, $i = 1, 2, \ldots, n$.

Applying Galerkin's principle, the following governing equation is obtained:

$$\left(\frac{\bar{r}}{(1-2\mu)(1+\mu)}\right)\sum_{j=1}^{n}\sum_{i=1}^{n} c_i \int_0^1 E(r)\left\{\begin{array}{l} \frac{\varphi_j\varphi_i}{(r\xi + a)} + \frac{\mu}{r}(\varphi'_j\varphi_i + \varphi_j\varphi'_i) \\ + \frac{(r\xi + a)}{r^2}\varphi'_j\varphi'_i \end{array}\right\} hd\xi = pah_0 c_j\varphi_j\Big|_0 \tag{10.11}$$

10.3 Results and discussions

The numerical values of E_0 and σ_{y0} for the cylinder material at the outer surface are taken as 210 GPa and 350 MPa, respectively, and results are evaluated at internal pressure of 100 MPa. The results are reported for grading parameter, β ranging from

2 to −2. The proposed methodology is validated in [11] and a very good agreement is reported. Following normalized variables are used:

$$\bar{r} = \frac{r-a}{b-a}, \bar{E} = \frac{E(r)}{E_0}, \ \bar{\sigma} = \frac{\sigma(r)}{p}, \ \bar{\sigma}_y = \frac{\sigma_y(r)}{\sigma_{y_0}}, \bar{u} = \frac{u\ E(r)}{b\ \mathrm{p}}, \ \bar{\sigma} = \frac{\sigma}{\sigma_{y_0}}$$

In Figure 10.2, Young's modulus variation for different grading parameter is plotted. It is evident from eq. (10.3) and Figure 10.2 that positive β value results in cylinder with increasing modulus along the radius while negative β ensures that the cylinder is stronger at the inner radius. The results are calculated for internally pressurized cylinders of aspect ratio (a/b) 0.3 and 0.5 at different grading parameter, β.

In Figure 10.3(a)–(c) the normalized displacement and stresses are plotted for cylinder with aspect ratio 0.5. In Figure 10.3(a) it is evident that for positive β, the displacement at inner radius is minimum and increases at locations away from the center, thus experiencing a "pull" in terms of material flow or displacement at the center of the cylinder. However for negative β, the displacement is maximum at the inner radius with decreasing magnitude at locations near the tip. This in general would result in a "push" to the material flow or displacement. The normalized radial and tangential stresses are shown in Figure 10.3(b) and 10.3(c), respectively. The effect of "pull/push" resulting from radial variation of displacement is evident on the radial and tangential stresses. In Figure 10.3(b)–(c), the results are compared with Chen and Lin [11] establishing the validity of the present methodology. Interestingly, the variation of displacement and stresses with β plotted in Figure 10.3 points toward the possibility of attaining constant displacement irrespective of radial location within

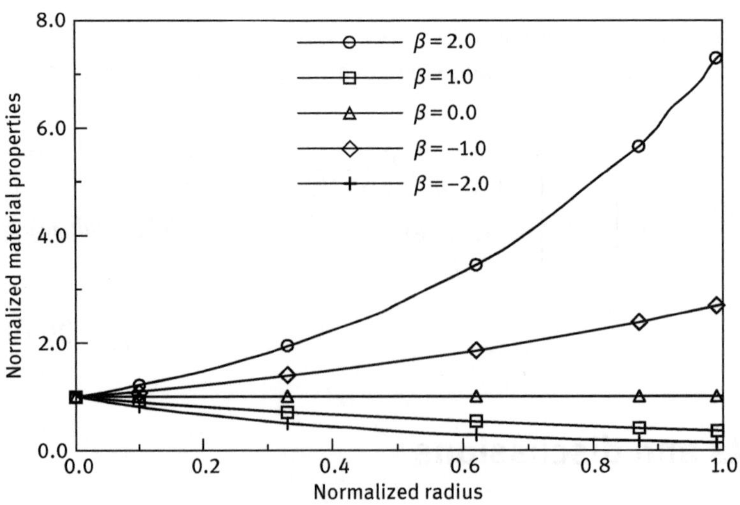

Figure 10.2: Variation of material properties for the FG cylinder.

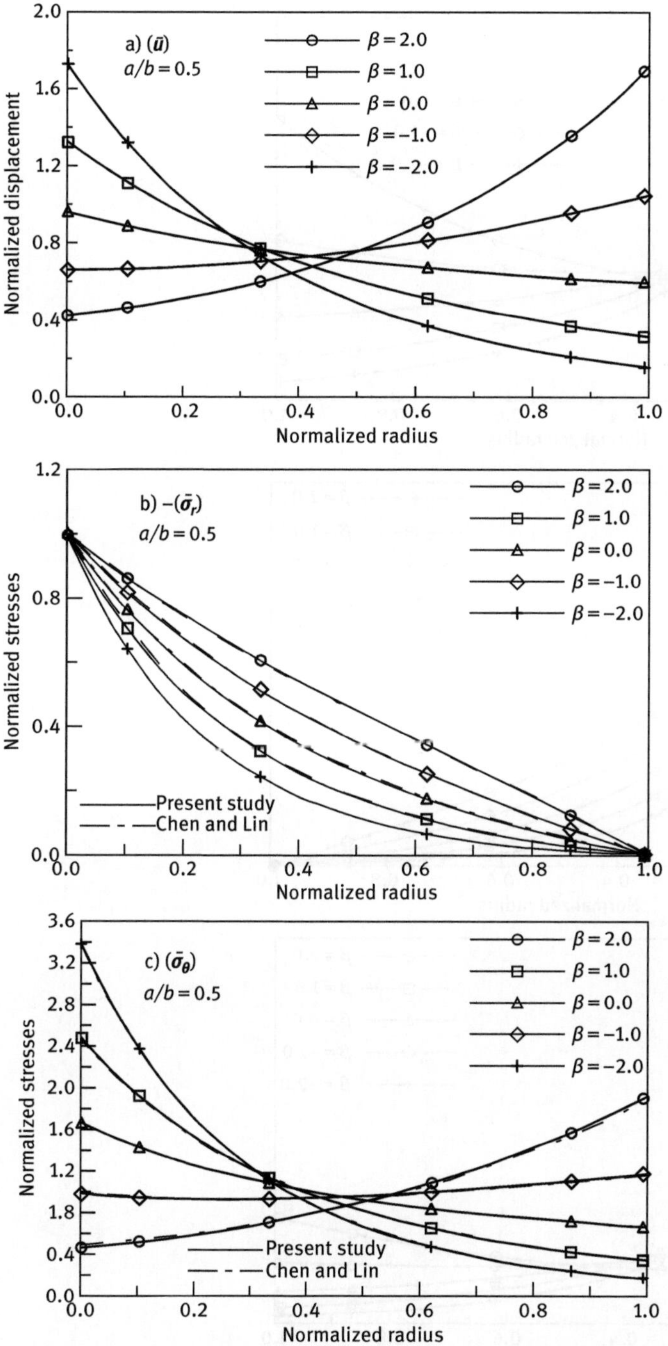

Figure 10.3: Normalized displacement, radial and tangential stress (a/b = 0.5).

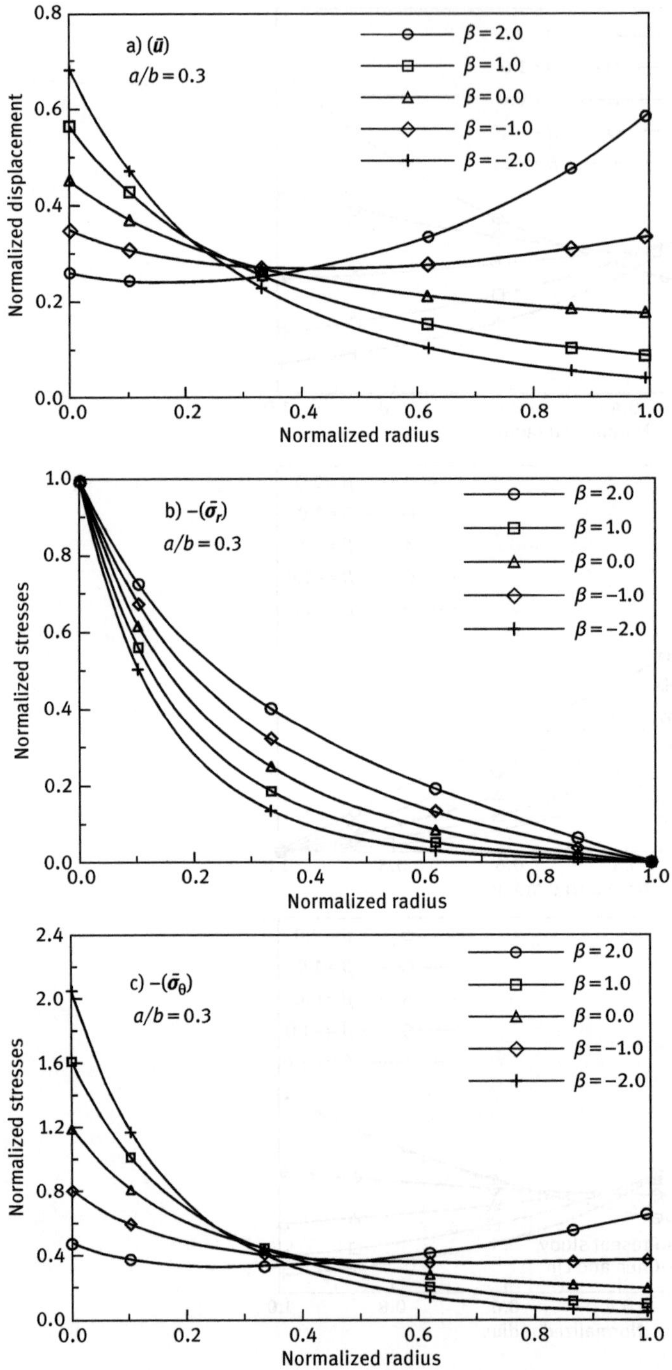

Figure 10.4: Normalized displacement, radial and tangential stress (a/b = 0.3).

the cylinder by careful tailoring of material at a certain β value that would result in uniform tangential stresses throughout the radial coordinate.

In Figure 10.4(a)–(c), similar results are plotted for cylinder with aspect ratio 0.3. The displacement and stresses of the cylinders give a good insight into the effects of grading parameter β on the performance of cylinder.

10.4 Conclusions

A numerical scheme is proposed to approximate the solution of deformation and stress distribution in a hollow cylinder subjected to internal pressure and is made of FGM. A strong concurrence with existing benchmarks established the validity of the proposed approximation. The influence of the gradient parameter β is studied on the stress and displacement field of the cylinder. The results obtained present a good insight into the variation of stress and deformation with grading parameter β and reveal the existence of grading parameter resulting in uniform strength cylinder.

References

[1] Tutuncu N, Ozturk M. Exact solution for stresses in functionally graded pressure vessels. Composites Part B: Eng 2001;32(8):683–6.

[2] Verijenko VE, Adali S, Tabakov PY. Stress distribution in continuously heterogeneous thick laminated pressure vessels. Compos Struct 2001;54:371–7.

[3] Tarn J-Q. Exact solutions for functionally graded anisotropic cylinders subjected to thermal and mechanical loads. Int J Solids Struct 2001;38:8189–206.

[4] Jabbari M, Sohrabpour S, Croccolo Eslami MR. Mechanical and thermal stresses in a functionally graded hollow cylinder due to radially symmetric loads. Int J Press Vessels Pip 2002;79(7):493–7.

[5] Shao ZS. Mechanical and thermal stresses of a functionally graded circular hollow cylinder with finite length. Int J Press Vessels Pip 2005;82:155–63. DOI:10.1016/j.ijpvp.2004.09.007

[6] Xiang H, Shi Z, Zhang T. Elastic analyses of heterogeneous hollow cylinders. Mech Res Commun 2006;33:681–91. DOI:10.1016/j.mechrescom.2006.01.005

[7] Tutuncu N. Stresses in thick-walled FGM cylinders with exponentially-varying properties. Eng Struct 2007;29:2032–5. DOI:10.1016/j.engstruct.2006.12.003

[8] Vullo V, Vivio F. Elastic stress analysis of non-linear variable thickness rotating disks subjected to thermal load and having variable density along the radius. Int J Solids Struct 2008;45: 5337–55. DOI:10.1016/j.ijsolstr.2008.05.018

[9] Peng XL, Li XF. Thermoelastic analysis of a cylindrical vessel of functionally graded materials. Int J Press Vessels Pip 2010;87:203–10. DOI:10.1016/j.ijpvp.2010.03.024

[10] Croccolo D, De Agostinis M. Analytical solution of stress and strain distributions in press fitted orthotropic cylinders. Int J Mech Sci 2013;71:21–9. DOI:10.1016/j.ijmecsci.2013.03.002

[11] Chen YZ, Lin XY. Elastic analysis for thick cylinders and spherical pressure vessels made of functionally graded materials. Comput Mater Sci 2008;44:581–7. DOI:10.1016/j.commatsci.2008.04.018

Index

ABAQUS 109
absorbers 57
absorption 143
absorptivity 64
acceleration 122
accelerometer 129, 144
acceptability 9
accuracy 24, 28, 29, 37, 42, 59, 61, 80, 121, 145
acetone 64
adaptive 94, 121
additive 37, 42, 43, 46, 50, 51, 59
adhesive 22
aerospace 1, 6, 12, 37, 50, 57, 119
agglomeration 31
agitating 26
aircraft 21, 22, 23, 28, 57, 86, 121
AISI 47, 61, 62, 63, 64, 70
algebraic 94, 159
algorithm 94
alloy 8, 10, 21, 22, 23, 29, 30, 31, 32, 33, 34, 35, 37, 44, 45, 47, 49, 61, 62, 63, 65
alloying 21, 22, 31, 48
alumina 7, 13, 46, 47, 108
aluminide 9
aluminium 21, 122
amalgam 22
AMC 36
AMCs 30, 32, 34, 37
amplitude 29, 122, 129
analysis 12, 29, 34, 45, 61, 66, 68, 69, 85, 90, 91, 100, 119, 146, 148, 156, 157
analytical 74, 76, 79, 90, 99, 100, 109, 112, 121, 122, 156
analyzer 146
anisotropic 156
ANN 76
annealing 23
annular 91
approximation 100, 126, 157
armour 3, 23
ASTM 59, 66, 144, 146
atmospheric 12
atomization 63
austenitic 63, 64, 67, 68, 70
automotive 28, 86
axial 32, 46, 60, 64
axisymmetric 83
ballistic 3, 57
bamboo 1, 56, 73, 82, 83, 85, 99
barrier 1, 12, 13, 41, 49, 56
behaviour 3, 6, 7, 8, 9, 29, 31, 32, 37, 121, 125
beverage 22
bidirectional 141
bimetal 10, 62, 63
binder 10, 59
biocompatibility 49
biodegradability 74
bioengineering 1
bioglass 46
biological 56, 57, 74
biomaterials 57, 75
biomedical 43, 45, 46, 50, 119
brazing 22
Brinell 8, 34
bromide 10
bronze 4

cantilever 122, 129, 131
capacitor 2, 6
carbide 8, 9, 10, 12, 30, 31, 47, 48, 58
carbohydrate 74
carbothermal 10
carburation 58
catalytic 11
categories 8, 55, 59, 74, 80
cavity 24, 25, 56, 143
cellular 45, 62
cellulose 85
centrifugal 8, 9, 34, 35, 42, 59, 80, 82, 120, 142
ceramic 1, 4, 5, 6, 7, 8, 9, 10, 12, 42, 45, 49, 56, 57, 58, 59, 62, 99, 100, 101, 120, 124
characterized 7, 47, 62, 63, 68, 69
chemically 66, 73, 77
chemist 22
chloride 10, 22, 66
chromium 22
circuits 10
circumference 8
CNT 36, 37
coaxial 46, 60, 64
cobalt 10, 11, 58
CoCrMo 45

https://doi.org/10.1515/9783110545104-011

coefficient 3, 29, 56, 91, 92, 94, 108, 119, 124
colloidal 7
combination 28, 125, 128, 158
combustion 9, 12, 28
component 2, 3, 4, 12, 26, 28, 30, 32, 37, 44, 46, 48, 50, 51, 55, 56, 57, 59, 60, 61, 62, 63, 64, 67, 74, 75, 77, 80, 85, 99, 101, 119, 121, 124, 142
components 25
composite 2, 21, 41, 55, 56, 61, 63, 66, 73, 74, 75, 76, 85, 99, 109, 119, 120, 121, 122, 140
composition 1, 2, 3, 7, 8, 10, 11, 21, 23, 36, 41, 42, 44, 48, 50, 55, 58, 63, 74, 75, 77, 80, 81, 82, 119, 148
compressive 34, 44, 45
computational 63, 131, 141
computerized 76, 77
concept 42, 55, 56, 57, 58, 73, 75, 77, 80, 122, 125, 129, 141
concurrence 163
conduction 91
configuration 44, 75
constitutive 29, 101, 158
constraints 80
contamination 66
conventional 4, 32, 42, 55, 59, 76, 77, 81, 83, 120, 125
corrosion 22, 35, 42, 119, 120
criteria 75, 90
crust 21
crystallization 68
crystallographic 22
Cu 5, 9, 21, 61, 63, 69
customized 42, 50
cylindrical 141, 156

dampers 12
decomposition 3, 11
deflection 112, 115, 122
deformation 3, 26, 28, 29, 31, 42, 99, 122, 141
degradation 74
delaminated 62
demonstration 81
demouldablity 7
dendrite 26, 68, 69
densification 3
density 5, 6, 22, 28, 29, 34, 35, 43, 45, 61, 62, 74, 81, 85, 100, 102, 121, 124, 129, 156
dental 56, 57
deposited 2, 7, 45, 46, 47, 48, 49, 60, 62, 65, 66, 67, 68, 69
deviation 94, 126
diameter 13, 25, 26, 28, 60, 65, 144
diamond 11
die casting 24, 49
dielectric 2, 6
diesel 57, 119
differential 90, 100, 102, 104, 105, 106, 108
diffraction 34
diffusivity 12
dilution 59, 61, 66, 67, 68
dimensional 11, 24, 31, 32, 42, 61, 82, 85, 100, 109, 121, 156, 157
dimensionless 93, 94, 109, 129, 157
direction 10, 11, 28, 43, 45, 48, 50, 58, 62, 66, 70, 80, 81, 85, 99, 109, 115, 116, 119
discharged 61
discrete 82
dislocation 13, 28
dispersant 7
dispersion 34, 66
displacement 31, 101, 105, 109, 116, 122, 156, 157, 158, 160, 163
dissipation 90
divergence 121
DMA 146, 148
DP 132, 133
dryer 142
ductility 22
durability 4, 75, 76, 141
duralumin 21
duration 109, 112
durometer 146
dynamically 122

economical 120
effectiveness 9
efficiency 12, 61, 91, 94, 122, 131, 140
eigenmatrix 108
eigenvalue 108, 121, 124, 128
eigenvector 124
elasticity 3, 6, 9, 44, 81, 121, 156
electrochemical 5, 11
electrocoating 6
electrode 6
electrolysis 22
electromagnetic 1, 26, 30, 31

electron 25, 32, 34, 66
electrophoretic 6
electrothermal 9
elemental 41, 63, 66
elevated 13, 29
elimination 8, 48, 56
elongation 31, 74
embedded 73
EMPA 13
enhancement 32
environmental 1, 74, 120, 131
environmentally 74
epitaxial 48, 63, 68, 69
equipment 38, 63
eradicate 56
erosion 76
ethanol 66
eutectic 26
evident 94, 95, 96, 147, 149, 150, 151, 152, 160
examination 25, 26
exceptional 35
exchangers 90
excitation 122, 129
exhaust 57
exothermic 9
expansion 3, 29, 49, 56, 76, 119, 120, 121
exploitations 73
exponential 108, 119, 156, 158
exposure 61
extrusion 23, 25, 26, 28, 59

fabrication 1, 22, 23, 30, 32, 33, 34, 35, 36, 37, 43, 44, 45, 55, 79, 82, 120, 142
FCG 29
FCI 29
FDM 37
feasibility 9, 44, 45, 62, 63
FEM 81, 83, 120, 121
ferrous 24
FG 99, 100, 101, 102, 103, 105, 109, 112, 115, 156, 157
FGCMs 75
FGM 1, 2, 3, 4, 5, 6, 7, 8, 9, 10, 11, 12, 13, 34, 35, 36, 37, 41, 42, 43, 44, 45, 46, 47, 48, 49, 50, 51, 55, 73, 74, 75, 76, 77, 79, 80, 81, 119, 120, 124, 128, 129, 131, 141, 142, 143, 144, 146, 147, 148, 149, 150, 156, 158
FGMMC 8
FGNF 80
FGP 121, 123, 127, 129
fibril 85
filaments 10
fin 90
flax 73, 82, 85, 86
flexibility 59
flow rate 10, 11, 47, 61, 65
flowability 61
fluid 7, 90, 121, 156
flux 32
foils 22
forging 8, 21, 22, 23, 26, 27
formation 3, 10, 21, 48, 49, 62
fraction 8, 9, 22, 28, 29, 30, 31, 32, 34, 36, 44, 45, 47, 56, 58, 62, 80, 85, 101, 119, 120, 121, 123, 156
FRF 121, 122, 125, 129, 131
friction 49, 59, 149
functionally 1, 34, 41, 42, 43, 55, 57, 58, 59, 73, 90, 99, 119, 140, 156
furnace 1, 23, 32, 144
fused 22, 37
fusion 43, 59, 119
fuzziness 124, 126

gaseous 1
gear 28, 49, 140
gel 6, 7
generation 13, 75, 79, 91
geometric 100, 120, 121, 122
geotextiles 86
governing 99, 100, 123, 158, 159
gradation 1, 2, 5, 50, 56, 58, 125, 144, 147
gradient 2, 3, 5, 8, 10, 11, 35, 37, 41, 43, 45, 46, 48, 55, 56, 58, 69, 76, 77, 80, 99, 120, 158
grain 7, 22, 25, 26, 28, 30, 33, 34, 48, 61, 64, 68, 69, 70
granular 26, 34
graphite 4, 11, 31, 32
gravitational 5, 8
gravity 21, 24, 25, 30, 59

HA 3, 7, 45, 46
hardness 1, 3, 5, 7, 8, 9, 10, 12, 13, 25, 29, 31, 32, 33, 34, 35, 36, 37, 48, 49, 62, 70, 75, 145, 146, 147
harmonic 116, 122
HDPE/HA 45

HDPE 45
hexagonal 34
hierarchy 76
homogenous 25, 148, 149
honeycomb 120
HSDT 99, 100, 101
HV 70
hybrid 6, 31, 32, 59, 119, 121
hydrochloric 66
hydroxyapatite 3, 12, 45
hyperbolic 91
hypersonic 41

ignition 9, 144, 145, 147
illustration 75, 77, 80
imbalance 3
impeller 31
implanted 50
Inconel 6, 46, 47, 48, 62, 63
indentation 47
industrial 6, 22, 42, 64, 75, 87, 91, 156
inert 59
inertia 105
infiltration 10, 30
infinitesimal 156
infrastructure 87
inhomogeneous 43, 56, 119
injection 13, 30, 140, 141, 142
innovation 1
insight 5, 95, 163
instigated 74
instill 51
insufficient 61
insulations 86, 119
integrated 75, 80, 121
intensiveness 131
interdisciplinary 75
interfacial 32, 38, 45, 58, 148
interlayer 46, 62
intermetallic 4, 8, 25, 49, 50, 62
interval 66, 121, 125, 126, 127, 128
invention 21, 44
inverse 9, 47, 61, 62, 96
investigation 3, 12, 13, 37, 45, 47, 49, 95, 144
ion 2, 4, 10
irradiation 61
isogeometric 100, 121
isotropic 95, 121

jute 73, 82, 85, 86

kinetic 9, 10, 102

laboratory 7, 64
laminate 36, 56, 121, 156
lamination 3, 43, 59
laser to create 59
laser 6, 10, 37, 43, 55
lattice 45
length 82, 92, 93, 94, 95, 96, 129, 157
lens 64
lignocellulose 82
lignocelluloses 74
linearity 100
linearly 62, 96
location 35, 129, 131, 141, 146, 147, 148, 160
longitudinal 62, 90, 91, 92, 94, 129
lubricating 49

machinability 21, 38, 77
macroscale 55
macrostructure 66
magnalium 21
magnetic 58, 75, 120
magnitude 108, 109, 160
manufacturer 63
martensite 70
mathematical 8, 90, 91, 125, 129, 157
MATLAB 94
maximum 4, 8, 9, 21, 32, 66, 94, 122, 128, 148, 149, 157, 160
measurement 29, 66, 147
mechanical 3, 5, 6, 7, 8, 9, 13, 22, 25, 26, 28, 29, 30, 31, 32, 33, 34, 36, 37, 44, 45, 46, 47, 48, 56, 61, 62, 63, 74, 75, 81, 85, 119, 120, 121, 140, 141, 146, 148, 156
mechanism 8, 10
mesh 31, 34, 83, 94, 129, 143, 144
metal 37
metallographic 26, 66
metallurgy 2, 8, 30, 33, 35, 36, 37, 42, 59, 82, 120
metal(s) 1, 4, 5, 6, 8, 10, 11, 12, 21, 22, 23, 24, 25, 26, 27, 28, 30, 31, 32, 42, 46, 55, 99, 100, 120, 124
methane 11
methodologies 75, 76

metrics 74
Mg 21, 25, 26, 28
MH 66
microfibrils 82, 85
micrograph 31, 34, 66
microhardness 47, 48, 49, 61, 62, 66, 70
microscope 25, 31, 32, 34, 66
microstructure 3, 11, 22, 23, 25, 26, 30, 31, 32, 33, 34, 35, 36, 37, 41, 44, 45, 47, 48, 55, 58, 61, 62, 63, 64, 66, 67, 75, 82, 85, 120
minimization 90, 94, 157
missile 119
mixture 2, 7, 8, 41, 85, 119
MMC 31, 32, 34
mobility 149
modification 4, 58, 122, 141
moisture 142
mold 49, 140, 141, 142
molecular 149
Molybdenum 63
monomer 7
morphology 4, 5, 11, 25, 34, 46, 61, 82
motor 142, 143
multifunctional 55, 74
multilayer 122, 156
multipass 46
multiphase 58
multistructures 46

nano 55
nanocomposite 141
nanocomposites 3, 44
nanocrystalline 47
nanofibers 3
nanomaterials 9
nanoparticle 32, 44
nanotube 36, 141
nanowire 11
neodymium 25
NFCs 73, 74, 76
NFs 74
Ni 9, 10, 21, 46, 49
NiCr 62
NiCrAl 13
NiCrAlY 12
Niobium 63
nitride 10, 30, 32, 34
Nm 151, 152, 153
nodal 124
noise 144
nominal 28, 120
non-hardenability 70
nonhomogeneous 2, 119
nonlinear 90, 100
normalized 93, 94, 109, 112, 115, 126, 158, 160
novel 41
nozzle 4, 57, 60, 61, 64
numerical 12, 45, 100, 115, 121, 127, 141, 157, 159
nylon 44, 140

one-dimensional 11, 100, 156
optical 25, 29, 31, 32, 58
optimal 9, 45, 49, 62
optoelectronic 11, 58
organic 5, 6
organisms 74
organization 43
orientation 22, 28
orthogonal 106, 124, 158
orthopaedic 57
orthotropic 101, 156
osteoblast 49
oxidation 10

palm 74, 86
parabolic 91, 93, 94, 95, 96
paradigm 12, 75
parameter 4, 5, 6, 8, 10, 12, 13, 44, 45, 47, 48, 49, 51, 62, 65, 75, 77, 90, 91, 94, 95, 96, 121, 122, 127, 128, 131, 140, 146, 157, 158, 159, 160, 163
particle 3, 4, 5, 6, 8, 10, 26, 30, 31, 32, 34, 35, 36, 37, 47, 48, 61, 63, 80, 121, 141
particulate 8, 9, 30, 31, 32, 34, 44
patent 4, 10, 11, 47
performance 3, 9, 32, 37, 56, 74, 75, 91, 94, 95, 96, 119, 140, 141, 143, 163
perimeter 91
perturbation 100
petrochemical 58
pharmaceuticals 22
phenolic 3
phenomenon 4, 7, 12
photo 59
photochemical 74
photovoltaic 87
physicomechanical 9

physics 51
piezoelectric 1, 7, 58, 121
piston 21, 57
plasma 2, 5, 10, 12, 13
plaster 8, 24
plasticity 27
plumber 144
poisson 81, 101, 108, 122, 124, 129
pollutant 64
polyamide 45, 141
Polycaprolactone 44
polycrystalline 11
polyester 5
polyethylene 45, 49
polyketone 141
polymerization 7, 59
polynomial 121
polypropylene 141
porosity 5, 7, 8, 9, 10, 33, 44, 45, 49, 56, 58, 61, 62, 91, 120
precipitate 22, 50
precision 8, 25, 28, 51, 122, 140
principle 4, 8, 9, 46, 102, 104, 122, 143, 157, 158, 159
probability 49, 122
production 2, 7, 21, 24, 25, 26, 28, 30, 32, 37, 42, 43, 46, 51, 55, 58, 59, 74
profile 66, 70, 91, 94, 96
projectile 3
pronounced 136
propagation 3, 6, 13, 121
propeller 119
proportion 142
proportional 31, 47
propulsion 57
prostheses 49
prosthesis 2
protein 85
pulp 86
pulse 100

quadratic 129
quantify 77, 122, 129
quenching 11

rackets 119
radial 83, 85, 91, 100, 157, 158, 160, 163
radiation 8, 91
radiative 9, 91
radius 147, 157, 158, 160
railing 22
ramming 24
rapid 11, 42, 48, 62, 70, 75, 80, 86
ratio 2, 10, 11, 23, 28, 47, 59, 65, 81, 101, 108, 122, 124, 129, 157, 160, 163
Rayleigh 122
reactors 119
recrystallization 12, 28
rectangular 91, 94, 95, 121
reduction 3, 10, 12, 25, 31, 34, 37, 56, 61, 62, 147, 151
refinement 69
reflectability 22
refractive 58
refractory 24, 25
reinforcement 6, 8, 30, 31, 32, 33, 34, 35, 36, 58, 74, 86, 149
relatively 4, 59
relevant 75, 157
reliability 77
remanufacturing 46, 75
remelting 23
renewable 73
replacements 57
resin 3
resistivity 5, 57, 76
resonance 122
response 42, 148
rigidity 115, 129
ripple 50
robot 64
rocket 57
rollers 26
rotation 142, 144
rotator 44, 142
roughness 44, 61, 62
rubber 149

scaffold 44, 45
scattered 25
schematic 1, 31
Schmidt 121, 159
scientific 1, 2, 4, 7, 9, 64
scrap 22, 32
sectional 26
sector 2
sedimentation 5, 59
segmentation 48

segregation 33, 35
selective 30, 43, 46
SEM 13, 66, 67
semiconductors 10
sensitive 60
sensor 1, 2, 144
shaft 32, 57, 119, 143, 144
shallow 61
shrinkage 33
silica 44
simulation 29, 37, 77
sintering 2, 3, 4, 7, 34, 36, 43
sinusoidal 105, 106, 149
sisal 73, 82, 85, 86
situ 8
skeptical 51
SLS 43, 44, 45
slug 28
slurry 7, 24
soldering 22
solidification 5, 8, 23, 26, 29, 33, 46, 61, 62, 63, 68
solution 6, 7, 42, 77, 90, 91, 94, 100, 109, 112, 131, 156
solvent 6
spacecraft 56, 119
spatial 1, 2, 44, 73, 91, 156
species 73
specimen 3, 11, 13, 32, 45, 62, 65, 66, 144, 145
spectrometer 25
spherical 46, 156
spheroid 33
spines 91
spiral 85
spur 140
squeezing 22
stability 13, 120, 121
stainless 10, 47, 48, 55
standardization 59
static 121, 122
stationary 108, 143
steeply 149
stiffness 9, 44, 83, 104, 107, 124, 125, 128, 140, 141, 149
stirrer 26, 32
stochastic 120, 121
strength 3, 7, 8, 9, 10, 12, 13, 21, 22, 26, 29, 31, 32, 34, 35, 36, 42, 44, 45, 47, 49, 56, 58, 74, 85, 120
structural 1, 3, 5, 9, 13, 22, 34, 41, 44, 66, 73, 75, 80, 86, 119, 120, 122, 126
substrate 2, 6, 7, 10, 12, 13, 46, 47, 48, 49, 59, 64, 65, 66, 67, 68, 69, 70
superalloy 46
supersonic 12
surge 12
surgery 37
symmetricity 129
synthesis 4, 9
synthetic 75

taper 32, 122
teeth 50, 120, 141, 150, 151
TEM 13
temperature 3, 4, 5, 6, 7, 8, 9, 10, 11, 12, 13, 25, 27, 28, 29, 32, 33, 34, 41, 56, 57, 75, 90, 91, 92, 94, 96, 119, 120, 121, 140, 141, 142, 143, 146, 148, 149, 150, 153
temperatures 29
tennis 119
tensile 9, 21, 32, 34, 35, 45, 47, 48, 56
tensor 83
textiles 86
texture 23
thermal 1, 3, 5, 11, 29, 41, 48, 49, 56, 57, 59, 63, 69, 76, 90, 91, 93, 119, 120, 121, 140, 156
thermodynamic 48, 63
thermomechanical 4, 28, 148, 156
thermoplastic 24, 142
thermosetting 142
thresholding 82
tissue 3, 44, 45, 50, 56, 57
torque 141, 143, 150
toughness 3, 9, 11, 13, 56, 74, 75
toxicity 44
transient 91, 99
transitional 10
transmission 34, 57, 140
transportation 86
transversal 85
triangular 100, 107, 109, 112, 115, 125, 126, 127
tribological 32, 37, 56
tungsten 10, 12, 58
turbine 10, 12, 57, 119
twins 13

ultrasonic 31
ultrasound 7
ultraviolet 74
uncertainty 120, 121, 122, 126
undamped 124, 125
unidirectional 30, 141
uniformly 47, 61, 65, 106, 109, 112
uniqueness 26
unmelted 47
unreinforced 140, 141
unstable 29
UTS 9

validate 90, 94, 129, 160
valve 129
vapour 2, 10, 59
variation 1, 3, 7, 36, 42, 55, 67, 70, 79, 80, 82, 85, 90, 91, 94, 96, 119, 120, 128, 129, 148, 156, 160
vibration 31, 99, 100, 116, 120, 121, 122, 144
Vickers 12, 13, 29, 66
viscoelastic 4, 121, 149
vortex 31

WC/Co 7, 58
weldability 77
welding 22, 59
wettability 32, 38, 49
whisker 30

yield 7, 26, 44, 45, 94, 95, 157

Bei Fragen zur Produktsicherheit wenden Sie sich bitte an:
If you have any questions regarding product safety,
please contact:

Walter de Gruyter GmbH
Genthiner Straße 13
10785 Berlin
productsafety@degruyterbrill.com